Sustainable Smart Composites

The book discusses cutting-edge technologies, advancements in processing techniques, and real-world applications of composite materials in diverse fields such as structural engineering, aerospace, automobiles, and defense. It further covers important topics such as the design and development of green composites and optimizing the mechanical properties of composite materials.

This book:

- Discusses the use of fiber-reinforced polymer matrix composite materials and artificial intelligence in the development of sustainable composites.
- Presents state-of-the-art methodologies such as fused filament 3D printing and stereolithography to produce composite materials.
- Concentrates on the diverse equipment employed to assess the behavior of materials, with emphasis on the use of the latest cutting-edge optimization techniques, such as artificial intelligence, to forecast the behavior of the materials.
- Introduces a range of material and operational parameters that impact strength to predict the mechanical properties of the developed materials.
- Showcases a comprehensive examination of the prospective utilization of composite materials in lieu of traditional substances for wind turbine blades.

It is primarily written for senior undergraduates, graduate students, and academic researchers in fields including industrial and production engineering, manufacturing engineering, materials science, mechanical engineering, aerospace engineering, and automobile engineering.

Advancing Science and Engineering through Artificial Intelligence, Machine Learning, and Mathematical Modeling

Innovation in Healthtech: A Roadmap for Empowering Healthcare
Edited by: Rakesh Kumar and Meenu Gupta

Sustainable Smart Composites: Technology and Applications
Edited by: Hiral Parikh and Piyush P. Gohil

Sustainable Smart Composites

Technology and Applications

Edited by
Hiral Parikh
Piyush P. Gohil

CRC Press
Taylor & Francis Group
Boca Raton London New York

CRC Press is an imprint of the
Taylor & Francis Group, an **informa** business

Designed cover image: Shutterstock

First edition published 2025
by CRC Press
2385 NW Executive Center Drive, Suite 320, Boca Raton FL 33431

and by CRC Press
4 Park Square, Milton Park, Abingdon, Oxon, OX14 4RN

CRC Press is an imprint of Taylor & Francis Group, LLC

ISBN: 978-1-032-86377-1 (hbk)
ISBN: 978-1-032-86378-8 (pbk)
ISBN: 978-1-003-52727-5 (ebk)

DOI: 10.1201/9781003527275

Typeset in Sabon
by SPi Technologies India Pvt Ltd (Straive)

Contents

About the editors

Hiral Parikh is an academician and researcher with over two decades of teaching experience in mechanical engineering. She currently holds the position of an Associate Professor and Ph.D. Program Coordinator at the School of Engineering and Technology, Navrachana University, Vadodara, Gujarat, India. Dr. Hiral has earned her Master of Engineering (M.E.) in Machine Design from Birla Vishvakarma Mahavidyalaya, Gujarat, and a Ph.D. from CSPIT, CHARUSAT, Changa, Gujarat, India. Her expertise lies in the domains of **Fiber Reinforced Polymer (FRP) Matrix Composites, Hybrid Composites, Bio-composites, Tribology of FRP Composites, and Composite Manufacturing and Characterization**. Dr. Hiral has around 50 research papers published in prestigious journals indexed in SCI and Scopus. She has also presented her research work at numerous national and international conferences and has authored five book chapters published by reputable publishers. Dr. Hiral's innovation, a composite structure heat pipe, was granted a patent by the Indian Patent Office in February 2023 and also received an Indian design registration, underscoring its unique design. Her scholarly impact is reflected in her **H-index of 9** and i10-index of 6. Dr. Hiral has delivered numerous expert lectures, seminars, and workshops, fostering learning and collaboration among students and professionals. She has secured grants from both private and government organizations to organize seminars, short-term training programs, and workshops. She is a reviewer for several high-impact factor and peer-reviewed journals. She is also a member of Tribology Society of India (TSI) and the Indian Society of Technical Education (ISTE).

Piyush P. Gohil is currently working as an Head & Associate Professor in the Mechanical Engineering Department at Faculty of Technology & Engineering, The M.S. University of Baroda, Vadodara, Gujarat. He is also holding the position of NCC Officer in 1 Guj. Compo. Technical Regiment Baroda Unit. In his previous service he was a Professor & Head Mech. Engg. Dept, CSPIT CHARUSAT Changa. He did his Ph.D. in August 2010 from SVNIT-Surat, Gujarat, India. Dr. Gohil's area of research includes **Design, Manufacturing, and Tribology of Composites**. He is a life member of ISTE, TSI, ISAMPE, and the FRP Institute, as well as an annual member of SAE. He has guided 10 Ph.D. students (with 3 currently pursuing their research) and supervised more than 70 M.E./M.Tech. dissertations. He has published more than 130 research papers in referred journals, national and international conferences. He has also authored 15 book chapters, edited 1 conference proceeding, and curated a thematic journal issue with reputed publishers. His research impact includes over 1,800 citations, an H index of 21, and an i10 index of 33. He has delivered more than 70 expert lectures in various faculty development programs and many expert lectures to the student community. He is acting as a reviewer in many national and international journals. He has completed two MODROB projects of AICTE, New Delhi, India. He has also received one research project grant from DST Purse, SERB Delhi, and two from GUJCOST, Gandhinagar. He was awarded travel grants as a Young Scientist from DST INDIA in March 2011 for Malaysia and under UGC Merged scheme in March 2015 for London. Dr. Gohil also acted as Chairperson of the Board of Studies (BOS) in Mechanical Engineering at M.S. University and as an external expert BOS member at BVM, CHARUSAT, and P.P. Savani University. He has also acted as Deputy Director of the Research & Consultancy Cell at M.S. University of Baroda from March 2019 to March 22 and is also a part of various national-level expert committees at AICTE and the NBA.

Contributors

Timileyin O. Akande
Department of Mechanical Engineering
First Technical University
Ibadan, Nigeria

Oluwaseyi Omotayo Alabi
Department of Mechanical Engineering
Lead City University
Ibadan, Nigeria

Arup R. Bhattacharyya
Department of Metallurgical Engineering and Material Sciences
IIT Bombay, Powai
Mumbai, India

Sivaji Chakravorti
School of Materials Science & Engineering
NIT Calicut
India

Vijaykumar Chaudhary
Chamos Matrusanstha Department of Mechanical Engineering
Chandubhai S. Patel Institute of Technology
Charotar University of Science & Technology
Changa, Gujarat, India

Viren Chunilall
School of Engineering
University of KwaZulu-Natal
Durban 4041, South Africa
and
Biorefinery Industry Development Facility (BIDF)
Council for Scientific and Industrial Research (CSIR)
Durban, South Africa

Prasanjit K. Dey
Department of Metallurgical Engineering and Material Sciences
IIT Bombay, Powai
Mumbai, India

Rituparna Duarah
Polymer and Petroleum Group, Materials Sciences and Technology Division
CSIR North-East Institute of Science and Technology
Jorhat, Assam, India

Ayotunde O. Fasina
Department of Mechanical Engineering
Lead City University
Ibadan, Nigeria

Oluwatoyin Joseph Gbadeyan
Department of Chemical Engineering
University of KwaZulu-Natal, South Africa
School of Engineering
University of KwaZulu-Natal
Durban 4041, South Africa

Jayaramudu Jarugala
Polymer and Functional Materials Division
CSIR-Indian Institute of Chemical Technology (IICT) Uppal Road
Tarnaka, Hyderabad, India

Albin Joseph
School of Materials Science & Engineering
NIT Calicut
India

Prabu Krishnasamy
School of Mechanical Engineering
Vellore Institute of Technology
Tamil Nadu, India

S. Senthil Kumar
Department of Mechanical Engineering
SRM Institute of Science and Technology
Tiruchirappalli, India

Mayand Malik
School of Mechanical and Materials Engineering
Indian Institute of Technology Mandi
Himachal Pradesh, India

Ipoteng Justice Mphahlele
School of Engineering
University of KwaZulu-Natal
Durban 4041, South Africa

Aravindh Murugan
Karpagam Academy of Higher Education
Tamil Nadu, India

Karthi Nadarajan
KPR Institute of Engineering and Technology
Tamil Nadu, India

Amos O. Olaleye
Department of Mechanical Engineering
First Technical University
Ibadan, Nigeria

Radhika Panickar
Materials Science and Engineering
Tuskegee University
Alabama, USA

Gaurang Patel
Chamos Matrusanstha Department of Mechanical Engineering
Chandubhai S. Patel Institute of Technology
Charotar University of Science & Technology
Changa, Gujarat, India

Shailesh Pathak
Department of Chemical Engineering
University of California
Santa Barbara, USA

G. Rajamurugan
School of Mechanical Engineering
Vellore Institute of Technology
Tamilnadu, India

Kawaljit Randhawa
Chamos Matrusanstha Department of Mechanical Engineering
Chandubhai S. Patel Institute of Technology
Charotar University of Science & Technology
Changa, Gujarat, India

S. Rathinavel
Department of Mechanical Engineering
SRM Institute of Science and Technology
Tiruchirappalli, India

Emmanuel Rotimi Sadiku
Department of Chemical, Metallurgical and Materials Engineering (Polymer Division)
Institute of NanoEngineering Research (INER)
Tshwane University of Technology
Pretoria, South Africa

Prateek Saxena
School of Mechanical and Materials Engineering
Indian Institute of Technology Mandi
Himachal Pradesh, India

T. S. Senthilkumar
Department of Mechanical Engineering
K. Ramakrishnan College of Technology
Tiruchirappalli, India

Vishakha Sharma
Department of Mechanical Engineering
University College of Engineering
Kota, India

Gokulkumar Sivanantham
KPR institute of Engineering and Technology
Tamil Nadu, India

C. B. Sobhan
School of Materials Science & Engineering
NIT Calicut
India

Shijo Thomas
School of Materials Science & Engineering
NIT Calicut
India

Astha Vijaivargia
Department of Mechanical Engineering
University College of Engineering
Kota, India

Gurupranes Sivaraj Vijaya
Dr. Mahalingam College of Engineering and Technology
Tamil Nadu, India

Chapter 1

Development of advanced composites reinforced with natural fibers

Prabu Krishnasamy and G. Rajamurugan

1.1 INTRODUCTION

The growing environmental concern, social concern, environmental restrictions, and energy consumption have prompted people to adopt environmentally friendly materials. Composites' carbon footprint and emissions of greenhouse gases have shrunk because of the increased usage of natural resources in their construction. Natural fibers are obtained from different plants, animals, and minerals. Plant-based natural fibers comprise cellulose, hemicellulose, and lignin. In recent years, country regulations have been stringent toward ecosystem and recyclability. In line with that natural fibers (NFs) reinforced composites are getting prioritized [1]. In general, NF composites are made up of epoxy resin [2] and a possible replacement for synthetic fibers to prepare eco-friendly materials [3]. Figure 1.1 shows the evolution of composite materials. Fabrication of natural fiber composite faced many challenges because natural fibers have a very high moisture absorption rate, and the raw fiber is not compatible with the matrix. Several problems must be resolved, and they also include low durability, low-impact strength, poor fire resistance, and poor interfacial adhesion between the matrix and natural fibers. Several studies have investigated natural fiber composites and their characteristics. Previous investigations have contributed to the development of different techniques for modifying natural fibers and resins to meet the growing need for environmentally friendly materials in various contexts. It is necessary to conduct a comprehensive analysis of green composites, including their physical and mechanical properties, sources and types of natural fibers and resins, methods of modification and processing, applications, life-cycle assessment, and other attributes.

1.2 NATURAL FIBERS AND THEIR COMPOSITES

Nature offers a vast array of materials that possess exceptional qualities, and natural fibers are among the best materials for different applications. Natural fibers are intriguing substances obtained from plants and animals.

DOI: 10.1201/9781003527275-1

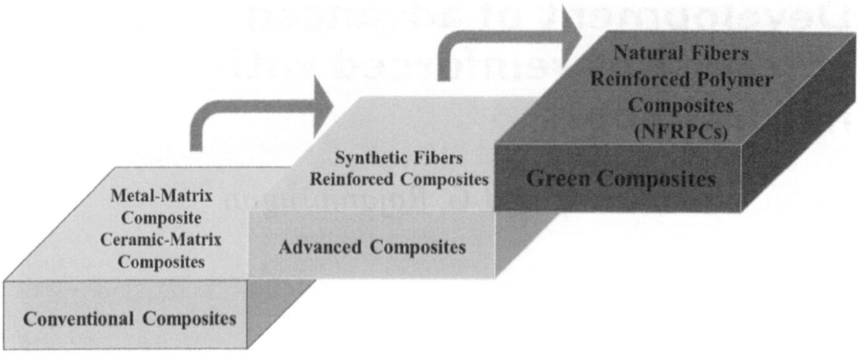

Figure 1.1 Composite material advancements [4].

(Open access.)

They have been employed for thousands of years in fabric, cords, and several other uses. Currently, there is renewed interest in advanced composites, which are emerging as a convincing substitute for conventional synthetic fibers. Natural fibers are a preferable alternative to synthetic fibers when constructing building components, particularly for lightweight applications [5]. Past findings show that fiber composites possess excellent robust mechanical properties, electrical resistance, acoustic insulation, effective thermal, and enhanced fracture resistance. Current uses encompass window frames, furniture, automotive components [6] packaging, pedestrian bridges, automobile interiors, shipping pallets, toys, roof tiles, composite furniture and construction elements. Recent research proves that advanced natural fiber composite materials offered better mechanical properties than challenging glass fiber (GF) composites. Recently, NF composites have replaced hardwood components in furniture and noise insulation. Also, it is used in panels, tubes, and sandwich plates. Fibers serve as reinforcements in composites, assuming many shapes such as chopped fibers, short fibers, long fibers, woven fibers, and yarns. Figure 1.2 shows the classification of natural fibers (NF).

1.3 COMPOSITE PREPARATION

1.3.1 Pre-treatment process (chemical treatment)

Recent developments in materials science have resulted in the replacement of materials composites reinforced with natural fibers [7]. The natural tendency of natural fibers to absorb water (hydrophilicity) can limit the performance of NF composites. To address water absorption of fibers, various chemical treatments can be applied to modify the fibers and improve their compatibility with polymer composite materials [8].

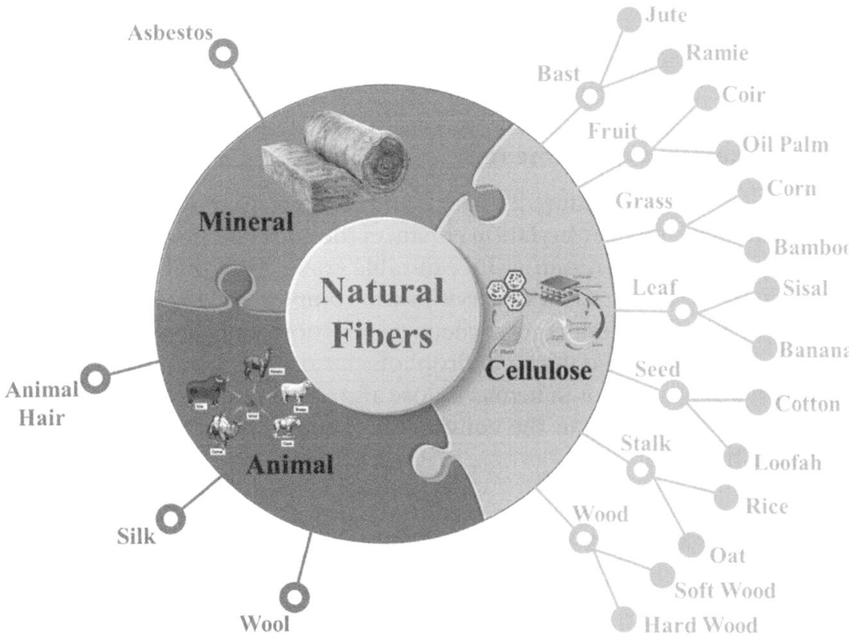

Figure 1.2 Classification of natural fibers.

1.3.1.1 Alkaline treatment

The alkaline treatment modifies the composition of organic fibers (OF), and sodium hydroxide (NaOH) is the chemical agent employed for this procedure [9]. Alkaline reagent is employed to modify the cellulose structure in plant fibers by a method known as alkalization, which involves cleaning the surface. The fibers are treated with NaOH at a temperature of 20°C for approximately 48 hrs. The fibers were then washed using H_2O (distilled water) and acetic acid to neutralize any remaining additional NaOH; the chemically treated fibers exhibited superior adherence to the resin, resulting in an elevation of surface energy and consequently improving the mechanical and thermal characteristics of the composites [10]. Alkalization has decreased the quantity of hemicellulose in the fiber, leading to improved mechanical properties compared to untreated fibers.

1.3.1.2 Silane treatment

The authors [11] reported that the silane groups function as a connector agent between the matrix and fiber, resulting in enhanced mechanical properties. The increase in strength is attributed to the creation of Si-OH groups, which establish robust connections with the –OH groups of the fibers. The hydrophobic silane undergoes polymerization and subsequently binds to the PMC by van der Waals forces. The success of low-density polyethylene composites lies in the excellent interface between the fibers and the polymer matrix [12].

1.3.1.3 Acetylation treatment

Acetylation is the procedure by which an acetyl group is added to the outer layer of natural fibers. Acetylation enhances the adhesion properties between the fiber and matrix, resulting in a durable connection that contributes to the favorable characteristics of NF-based composites [13]. The hydroxyl groups of the fibers undergo a chemical reaction with the acetyl groups, resulting in an increase in the hydrophobicity of the fibers. In general, the hydroxyl groups present in hemicellulose and lignin undergo a reaction with acetyl groups, resulting in the conversion of these compounds into hydrophobic substances.

1.3.1.4 Hydrogen peroxide treatment

The peroxides underwent decomposition to generate free radicals. After the alkali treatment, fibers are soaked in a 6% benzoyl/dicumyl peroxide-acetone solution for 30 min. Peroxides generate free radicals that react with cellulose and the future polymer matrix [14].

1.3.1.5 Benzoylation treatment

Benzoylation is employed to reduce the hydrophilicity of the fibers. This treatment enhances the adhesion between the matrix and fiber, hence augmenting the strength of the composites. In the process of benzoylation, the fibers undergo an initial treatment with NaOH, followed by a 15-min treatment with benzoyl chloride (C_6H_5COCl). Subsequently, the fibers were separated and subjected to 1-min ethanol treatment, followed by rinsing with H_2O (distilled water) and drying in an oven at 80°C for 24 hrs [15].

1.3.1.6 Potassium permanganate treatment ($KMnO_4$)

$KMnO_4$ serves as a chemical reagent to alter the interface between the matrix and fiber. The fibers treated with alkaline were then treated with potassium permanganate at various concentrations (ranging from 0.005% to 0.205%) for 1 min. Subsequently, the fibers were dried using an oven [16].

1.3.1.7 Stearic acid treatment

The non-woven jute fibers were soaked in varying concentrations of stearic acid in anhydrous ethanol for duration ranging from 1 min to 4 hrs, and they were dried at 100°C for 1 hr [17]. Short Sansevieria fibers were treated with

1% stearic acid dissolved in ethyl alcohol solution and combined with alkali-treated fibers in a steel vessel. The mixture is then stirred, and subsequently, the fibers were subjected to a drying process at 80°C for 45 min [18].

Table 1.1 shows variation in the chemical component that has been the subject of past research. There are a number of factors that disturb the progress of NF: the method of extraction, the timing of harvesting, and the technique of harvesting the main component, cellulose. A fiber's tensile strength is directly proportional to its crystallinity, thus crystalline cellulose in particular has a significant impact on this parameter. Since cellulosic plant fibers typically swell when exposed to water, they possess poor dimensional stability and high moisture absorption capacity [31]. Natural fiber strength is affected by cellulose, and UV degradation and char formation are both improved by lignin. Cellulose crystallinity is directly proportional to the amount of hemicellulose present. In the pulp and paper industry, lignin is typically removed during pulping because it is an undesirable part of the raw materials.

1.3.2 Post process (composite fabrication)

Several manufacturing techniques are available for composites, such as infusion, film stacking, lay-up, molding, pultrusion, and winding. The selection of an appropriate technique is contingent upon the characteristics of the initial materials, the desired shape of the finished result, and other financial considerations.

1.3.2.1 Hand layup

In general hand layup method was popular among the researchers. The surface of the mold is then covered with pre-cut reinforcement materials, such as fiber, fillers, wire mesh, etc. Air bubbles are removed from the reinforcement by applying and smoothing liquid resin using rollers. Repetition of this layering of reinforcement and resin process yields the necessary thickness and strength. Later, depending on the resin system, the laminate is either heated or allowed to cure at room temperature. Once it has cured, the composite component can be removed from the mold and sized to the desired shape and size. Figure 1.3 illustrates the different types of fabrication methods.

1.3.2.2 Compression molding

Compression molding is a rapid and efficient molding process that requires significant pressure. This technique involves the use of two coordinated metal molds to produce a composite product. In pressure decay, the bottom plate is stationary while the top plate is movable. The fortification and lattice are formed into a metallic shape and then securely held together among the reduction in pressure. Manufacturing process that utilizes both high

Table 1.1 Chemical and physical composition of various natural fibers

Type of fiber	*Cellulose (%)*	*Lignin (%)*	*Wax (%)*	*Pectin (%)*	*Hemicellulose (%)*	*Ash (%)*	*Silica (%)*
Rice	28–57 [19]	12–19 [19]	8–38 [15]	—	23–33 [19]	15–20 [19]	9–14 [19]
Ost	31–48 [19]	16–19 [19]	—	—	27–38 [19]	6–8 [19]	4–6.5 [19]
Wheat	29–51 [19]	16–21 [19]	—	—	15–32 [19]	4.5–9 [19]	3–7 [19]
Barley	31–45 [19]	14–15 [19]	—	—	24–29 [19]	5–7 [19]	3–6 [19]
Sida mysorensis	53.36 [20]	9.46 [20]	0.86 [20]	—	15.23 [20]	3.33 [20]	—
Momordica charantia	61.2 [21]	4.8 [21]	1.1 [21]	—	17.3 [21]	2.24 [21]	—
Sugar	32–48 [19]	19–24 [19]	—	—	27–32 [19]	1.5–5 [19]	0.7–3.5 [19]
Rye	33–50 [19]	16–19 [19]	—	—	27–30 [19]	2–5 [19]	0.5–4 [19]
Citrullus lanatus	53.7 [21]	10.1 [21]	3.2 [21]	—	12.5 [21]	2.17 [21]	—
Bamboo	26–73.83 [16]	10.15–36.88 [16]	—	—	12.49–31 [15]	1.7–5 [19]	0.7–3.51 [19]
Congon grass	34.23 [22]	26.99 [22]	—	—	36.23 [22]	—	—
Sabai	—	22 [19]	—	—	24 [19]	6 [19]	—
Cortaderia selloana grass	53.7 [23]	10.32 [23]	3.1 [23]	—	14.43 [23]	4.2 [23]	—
Cyperus pangorei	68.5 [18]	17.88 [18]	0.17 [24]		—	3.56 [20]	—
Phragmites communis	44–46 [19]	22–24 [19]	—	—	20 [19]	3 [19]	2 [19]
Esparto	33–38 [19]	17–19 [19]	—	—	27–32 [19]	6–8 [19]	—
Seed flax	43–47 [19]	21–23 [19]	—	—	24–26 [19]	5 [19]	—
Flax	71	2.2 [20]	1.5 [20]	2.3 [25]	18.6–20.6 [20]	—	—
Kenaf	31–57 [19]	9–21.2 [19]	—	—	20.3–33.9 [16]	2–5 [19]	—
Jute	45–71.6 [15]	21–26 [19]	0.5 [15]	0.2 [39]	13.6–21 [19]	0.5–2 [19]	—
Sun Hemp	41–48 [16]	22.7 [16]	—	—	8.3–13 [16]	—	—
Hemp	57–77 [19]	3.7–13 [19]	0.8 [26]	0.9 [25]	14–22.4 [19]	0.8 [19]	—

Sorghum bagasse	42.36 [11]	24.98 [11]	—	—	30.67 [11]	—	—
Ramie	68.6–91	0.6–0.7 [20]	0.3 [15]	1.9 [25]	5–16.7 [19]	—	—
Pineapple	81 [15]	12.7 [15]	—	—	—	—	—
Bagasse	55.2 [15]	25.3 [15]	—	—	16.8 [15]	—	—
Banana	60–65 [16]	5–10 [16]	—	—	19 [16]	—	—
Kenaf	37–49 [19]	15–21 [39]		—	18–24 [25]	2–4 [25]	—
Jute	41–48 [19]	21–24 [19]	—	—	18–24 [19]	0.8 [19]	—
Abaca	56–63 [15]	7–9 [16]	3 [15]	—	15–25 [16]	3 [19]	—
Sisal	47–78 [19]	7–11 [25]	2 [15]	10 [25]	10–24 [15]	0.6–1 [19]	—
Curaua	73.6 [15]	7.5 [15]	—	—	9.9 [15]	—	—
Water Hyacinth	65.07 [28]	11.38 [28]	—	—	15.07 [15]	—	—
Henequen	77.6 [25]	13.1 [25]	—	—	4–8 [25]	—	—
Areca palm leaf stalk	57.49 [29]	7.27 [29]	—	—	18.34 [29]	0.71 [29]	—
Sansevieria	79.7 [24]	10.13 [24]	0.86 [24]	—	10.15 [24]	1.4 [24]	—
Agave Americana C. fiber	68.54 [30]	6.08 [30]	—	—	18.41 [30]	3.29 [30]	0.56 [30]
Cotton linter	90–95 [19]	0.7–1.6 [19]	—	—	1–3 [19]	0.8–2 [19]	—
Cotton	82.7–90 [16]	—	0.6 [15]	—	5.7 [16]	—	—
Coir	32–43 [16]	40–45 [16]	—	—	0.15–0.25 [16]	—	—
Kapok	64 [15]	—	—	—	23 [15]	—	—
Rice husk	35–45 [15]	20 [15]	14–17 [15]	—	19–25 [15]	—	—

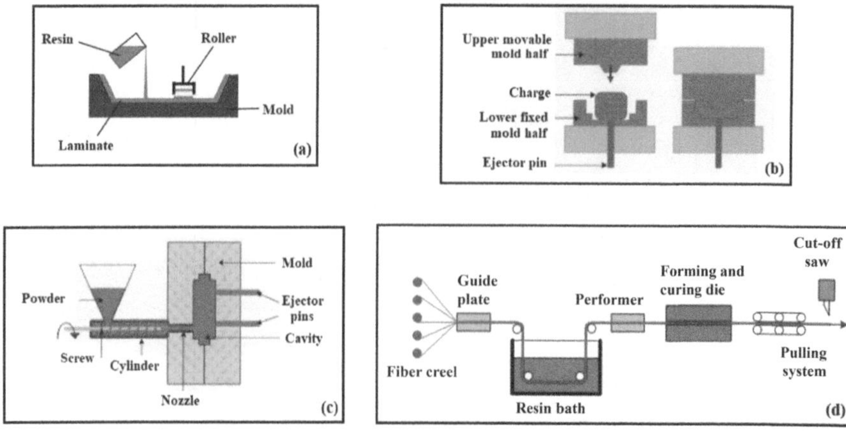

Figure 1.3 Different types of fabrication methods (a) hand layup method, (b) compression molding, (c) injection molding, (d) pultrusion method.

pressure and heat to shape items [32]. The application of pressure induces the material to completely occupy all the small spaces and crevices within the mold cavity, while the application of heat initiates a chemical reaction in the resin, causing it to solidify permanently. After the curing process, the mold is opened, exposing the completed component. This technology is particularly effective in large-scale production since it has quick cycle times and achieves great dimensional precision.

1.3.2.3 Injection molding

Injection molding is a prevalent method used to manufacture plastic components. It functions similarly to a sophisticated syringe. Plastic pellets are introduced into a heated barrel where they undergo the process of melting. A spinning screw agitates and propels the liquefied plastic toward a mold cavity that duplicates the intended ultimate form. After a sufficient amount of plastic is placed into the mold, the pressure is consistently maintained for a specific duration to guarantee effective solidification. Eventually, the mold is cooled and opened, enabling the accurately formed plastic component to be expelled. This technology is highly proficient in efficiently manufacturing delicate and complex parts with a remarkable level of uniformity [33].

1.3.2.4 Pultrusion method

Pultrusion is a continuous manufacturing process that produces consistent-shaped profiles made of fiberglass or other fiber-reinforced plastic (FRP). Visualize a mechanized transport system propelling filaments of fiber through a heated shaping apparatus. The fibers, often made of fiber glass or carbon, are initially impregnated with a liquid resin, such as epoxy or polyester. This resin functions as an adhesive, effectively joining the fibers together. Subsequently, the fibers and resin are drawn through a heated die, which molds them into the required cross-sectional shape, such as a rod, bar, or tube. The application of heat causes the resin to undergo a curing process [34] resulting in its solidification and permanent adhesion to the fibers. Once the FRP profile has been cured, it is released from the die and trimmed to the desired length.

1.4 PERFORMANCE AND CHARACTERIZATION

Fiber-reinforced composites consist of two distinct phases: the matrix and the reinforcement. Wielage et al. [35] classified low-density polyethylene (LDPE), polypropylene (PP), high-density polyethylene (HDPE), and polyether ether ketone (PEEK) as polymer plastic matrices. Paul et al. [36] and Rouison et al. [37] identified glass and carbon fibers as examples of ordinary synthetic fibers. In general, plastics have lower mechanical qualities, but their mechanical performance can be enhanced by including high-strength materials (fibers) as reinforcing media. Composite materials have been traditionally utilized in numerous commercial applications throughout history. As an illustration, ancient people utilized bamboo shoot particles in mud walls, whereas Egyptians in 1500 BCE employed them in laminated bonded wood [38]. The demand for composite materials, which offer great strength and are environmentally benign, is driving significant advancements in design, manufacturing techniques, and current technologies [39]. Natural fiber-reinforced polymeric materials are gaining traction as a potential replacement for traditional fibers, attracting the attention of researchers and engineers. The numerous benefits of NFs, such as their ease of decomposition, environmental friendliness, low cost, and lightweight nature, contribute to their popularity. The plants that produce NFs are classified as primary and secondary plants according to their utilization. Jute, sisal, and kenaf are cultivated as primary or important plants for their fiber content. On the other hand, coir, oil palm, and pineapple are examples of auxiliary or secondary plants that produce fibers or strands [40].

1.4.1 Mechanical properties of NFCs

Natural fiber-reinforced composites offer several advantages, such as cost reduction, weight reduction, and sustainable solutions. These properties are particularly advantageous in industries such as plastics, electronics, packaging, and automotive. In consumer applications, NFRC is used for domestic tables, chairs, interior paneling, and window panels [41]. Additionally, these materials can serve as a viable and environmentally friendly choice for use in autos and the inside paneling of airplanes. Table 1.1 presents the key characteristics of NFs as documented in the literature.

Table 1.2 Natural fibers mechanical performance [44]

Composites	*Cellulose (wt%)*	*Impact strength (kJ/m^2)*	*Flexural strength (MPa)*	*Tensile strength (MPa)*
Coir	32–43	26.43	46.63	23.68
Sisal	66–78	56.7	52.8	37.40
Jute	61.1–71.5	65.00	55.8	43.00
Banana	63–64	149.66	76.53	59.00
Baggase	—	110.66	56.7	42.40
Flax	71	191.71	75.40	59.85
Areca	—	93.33	25.00	27.50
Ramie	68.6–76.2	105.40	110.00	90.00
Lantana camara	—	32.30	55.45	19.08
Pseudo stem Banana	—	92.66	73.58	45.57
Groundnut shell	31.3	24.17	28.00	18.09
Rice husk	35.7	26.00	29.00	23.00
Coir pith	27.1	18.67	23.00	9.00
Banana + sisal-epoxy	63–64	98.66	62.00	25.00
Luffa + Groundnut-epoxy	31.3	27.33	58.95	39.31

The literature has multiple investigations that confirm the exceptional mechanical capabilities of these materials, as well as their limitations in terms of practical use. Venkateshwaran and Elaya Perumal [33] discovered that nanofillers decreased the substantial mass of composites. Bisaria et al. [42] investigated the mechanical properties of composites by varying the length of the fibers. They discovered that composites with a length of 15 mm exhibited the highest flexural and tensile properties. A study conducted by Kumre et al. [43] discovered that the use of NFs as reinforcement can effectively reduce tool tear and wear in the machining process. The polymer incorporates a filler made from wood and shell flour to enhance the composites' properties, including reducing shrinkage and improving resistance to deformation after molding. Table 1.2 illustrates the mechanical performance of NF.

The X-ray diffraction performance of natural fiber of different natural fibers is presented in Table 1.3. Also, it includes the crystalline intensity, amorphous intensity, crystallite size, and percentage of crystallinity index.

1.4.2 Vibration performance of NFCs

Researchers analyzed the impact of fiber volume and fiber arrangement on the vibrational characteristics of sisal and cotton fiber-reinforced polyester composites [50]. Composite properties showed a considerable improvement at 40% fiber loading. The sisal fiber percentage of the materials is 30%,

Table 1.3 XRD for natural fibers

Natural fiber	*Crystalline intensity at 2θ (°)*	*Amorphous intensity at 2θ (°)*	*Crystallinity index (CI) in percentage*	*Crystallite size (CS) in nm*	*References*
Flax	22	18	70°	2.8	[45]
Ramie	22	18	58°	—	[45]
Sisal	26	18	75°	—	[45]
Coconut	22.5	—	68°	—	[46]
Hemp	22	16.6	87.9°	—	[47]
Jute	22.5	18.5	58.90°	—	[48]
Borassus	22	16	38.4°	—	[48]
Spanish	22	18	52°	—	[48]
Nelumbo nucifera	22.2	16.3	—	—	[48]
Cornstalk	22	18	74°	3.8	[48]
Raffia textilis	22.5	18.5	64°	32	[48]
Cotton stalks	22	18	65°	5.5	[48]
Pineapple	22	18	58.6°	—	[48]
Luffa cylindrical	22	18	59.1°	—	[48]
Chakshir	22.2	16.8	48°	1.6	[49]

which increases their inherent frequencies by 34–300 Hz. The fiber concentration is also said to increase the materials' natural frequency. The vibrational behavior of hybrid composites made of Sansevieria cylindrica and coconut sheath was studied by Bennet et al. [51]. In all three modes, it was found that the hybrid composite of coconut sheath, sansevieria, and coconut sheet had a high natural frequency. The chemical treatment also affected the damping characteristics, which led to a rise in the natural frequency for CSC composites from 28.53 to 48.25 Hz. The damping ratio followed a similar pattern, rising from 0.0502 to 0.0612. The free vibration behavior of polyester composites made of short sisal fiber and banana fiber was further investigated by Senthil Kumar et al. [52]. According to their findings, the damping values are directly proportional to the fiber loading. Fifty percent fiber content and three fiber lengths improve SFPC's mechanical characteristics. A 50 wt.% fiber loading produced the best mechanical performance for BFPC with a 4 mm fiber length. Based on their investigation into luffa fiber's vibrational properties, Kalusuraman et al. [53] found that chemical treatment enhanced the mechanical and damping abilities of the composites. Once treated with NaOH, the luffa composites (which include 40% fiber) exhibited a natural frequency rise from 67 to 110 Hz. The natural frequency of the different composites is shown in Figure 1.4 by mode-I and damping ratio. Using the maximum values found in the mode for each composite, it is calculated. There is a higher natural frequency in mode I in polyester composites with a coconut/banana/coconut sheath. The natural frequency

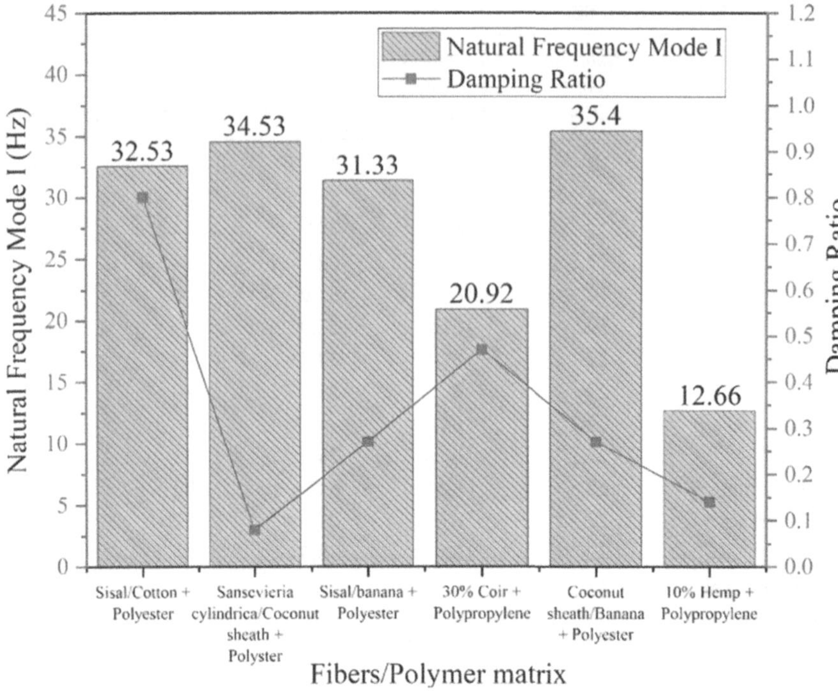

Figure 1.4 Comparative analysis of the natural frequencies of different natural fiber composites in mode-I and damping ratio.

of 10% hemp polypropylene composites was found to be 12.66 Hz, which is lower. Lamination increases the natural frequency compared to a single layer or fiber. The most extreme results reported in the published studies when operating in mode-I (bending) are used to illustrate it. Although there are damping ratio figures for several treated fibers, they are calculated from untreated fibers so that the data may be more easily understood. Composites of CCC (Coconut sheath/Coconut sheath/Coconut sheath) polyester had a damping value of 0.80234. When it comes to industrial vibrations, the CCC composite can handle more of them. Properties of coir-reinforced polypropylene composites relate to vibration damping [54]. The damping qualities of composites were found to be reduced when the weight percent (0%–30%) and fiber content were varied in this investigation. Compared to the other modes, the 30 wt.% composite had higher natural frequencies [55].

Compared to the post-buckled part, the pre-buckled section had a lower natural frequency. Weaved coconut sheath and short banana/polyester hybrid composite exhibited vibration behavior under untreated and treated (alkali) circumstances, as influenced by stacking sequence [56]. Both the CBC layered pattern and the treated fiber increase the suspension capacity. Using loadings ranging from 0% to 60% wt.%, with 10% increments in between, vibration experiments were conducted on noil hemp fiber-polypropylene composites [57]. Additionally, it looked into how the coupling agent affected vibration patterns. For composites containing 30-weight percent noil hemp, the stated damping ratio is the maximum value. The inclusion of POE-MAH and PP-MAH improved the storage modulus of composites reinforced with 30 wt.% noil hemp. Short bamboo/polyester composites with different fiber concentrations (15 wt.% and 10 wt.%) were the subjects of vibration experiments carried out by Jena [58]. Due to the increase in fiber content, the composite stiffness increased, which enhances the natural frequency [59].

Fibers of different dimensional lengths (5, 10, 15, and 20 mm) and a 30% fiber content were hand-lay-up produced into the composites. The storage modulus and loss modulus of the composites were found to be high when the fiber length was 15 mm. When looking at the vibration parameters of composites reinforced with flax fibers, Prabhakaran et al. [60] found that the attenuation was 51% higher in the flax fiber composite compared to the glass fiber composites. Natural fiber composites also have better damping capabilities. The effect of fiber orientation on damping is shown to be more substantial than that of fiber loading. Munde et al. [61] conducted an investigation into the vibration attenuation of polypropylene composites reinforced with sisal fibers. The composites were produced with varying fiber loadings (10%, 20%, and 30%). The optimal natural frequency and minimal damping are demonstrated by the 30% fiber loading.

1.4.3 Wear properties of NFCs

Research on hybrid composites made from natural fibers has shown that composites with fibers oriented in their normal direction have better wear and friction qualities compared to those with treated fibers. The most effective fiber alignment for reducing wear and friction was determined to be in the opposite direction of the sliding motion, irrespective of any treatment used. The wear performance of a material is affected by various factors, including the volume percentage of fibers, the applied load, and the compatibility between the fibers and the matrix. Typically, an escalation in the applied stress results in an elevated wear rate [62].

The wear behavior of both untreated and treated oil palm fiber-reinforced polyester composites was analysed [63]. The results revealed that the treated oil palm fiber composites, showed better wear resistance than the untreated fiber composite.

The wear characteristics of composites are affected by the formation of a coating on the surface of the counter material during sliding. Composites generally demonstrate a high coefficient of friction at the beginning of the running stage, which is then followed by a stable state, demonstrating the durability of the rubbed surface. On the other hand, the friction coefficient lowers during the steady state in comparison to the running condition because a smooth and more stable coating forms on the surface that opposes it.

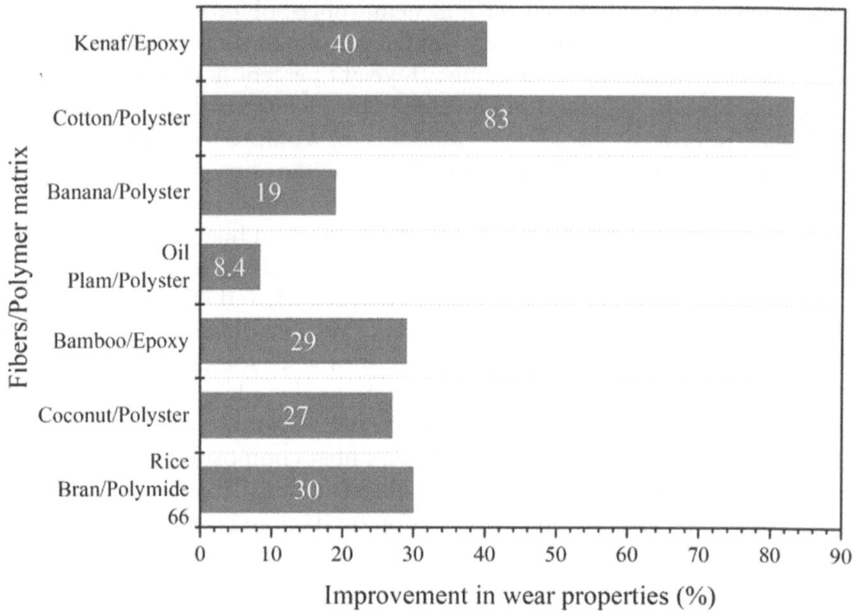

Figure 1.5 Enhancements in the wear properties are achieved through the utilization of natural fibers.

Wear can greatly diminish the performance of a material by modifying its composition and physical structure. Research has demonstrated that incorporating natural fibers into composites can frequently improve their ability to withstand wear and increase their efficiency in terms of friction. These fibers enhance durability and rigidity, resulting in enhanced resistance to wear. In many instances, the incorporation of fiber reinforcements has led to an almost complete improvement of approximately 100% [64]. Figure 1.5 provides a concise summary of the beneficial effects of natural fiber composite reinforcements on wear and friction.

1.4.4 Electromagnetic interference (EMI) shield properties of NFCs

Several studies have examined the EMI shielding capabilities of conventional substances such as metal, carbon, ceramics, polymers, and related composites. Carbon fiber has been opted for as the most popular material to produce EMI shielding material with polymer. The incorporation of fillers, such as graphene, carbon nanotubes, carbon black, and silver, would improve the EMI shielding efficiency. It was concluded that graphene is an important substance due to its excellent electrical conductivity. The graphene composite showed 21 dB of EMI shield effectiveness in the X-band frequency, whereas the carbon fiber reinforced graphene composite showed an EMI shield effectiveness of around 70 dB. However, adding NiP to Kevlar fibers increased EMI shielding efficiency from 13 to 19 dB, whereas adding NiPCu on carbon fiber increased the EMI SE by more than 76 dB. The CNT addition develops conductive pathways that improve EMI shielding efficacy in carbon/glass fiber composites (80 dB). Meanwhile, the use of synthetic fibers and their composites in EMI shielding remains a challenge because of their capacity for recycling and high expense. However, natural fibers are capable of replacing synthetic fibers because they can be renewable, biodegradable, friendly to the earth, and inexpensive, which can replace synthetic fibers.

Also, it was reported that the electrical conductivity of the material has significant EMI shielding properties. Therefore, natural fibers are non-conductive substrates and using them as an EMI shielding material is challenging. But the electrical conductivity of natural fiber can be enhanced through spray, dip, and electroless coatings. Among the techniques, electroless deposition is a viable method for uniform NiP deposition on a non-conductive natural fiber substrate. Many researchers have demonstrated that electrical conductivity improved the EMI shielding performance of fiber-reinforced composite materials. Also, a few studies supported that matrix type improved the electrical characteristics of composites. However, it covers the conductive tracks, rendering it difficult to estimate the process's effectiveness.

The effect of NiP-coated hemp fiber on the EMI shielding effectiveness was studied. The consequences of two-layers of coated hemp significantly improved the EMI shielding properties. It has been reported that the coated natural fiber composite showed better results than the base samples [65].

1.5 APPLICATIONS OF NATURAL FIBER COMPOSITES

The automotive and aerospace sectors have been actively producing various interior components using NF [66]. NF is utilized in the production of insulation materials for many purposes, including pouring insulation, blowing insulation, thermal insulation, ceiling panels, impact sound insulation materials, and acoustic soundproofing. NF offers a promising prospect for sustainable design, providing a wide range of construction materials, forms, and even enhancing existing regularly employed materials. Substituting NF for synthetic fibers in the realm of architecture is a viable option. It serves as a material for walling, sunscreens, cladding, and flooring. Mercedes-Benz now incorporates NF, including flax, wool, hemp, and sisal, in its components [67]. The composites reinforced with coir and polyester were utilized in various applications such as mirror casings, voltage stabilizer covers, paperweights, helmets, projector covers, and roofs. The boat GreenBente24 utilized flax fibers. Cotton, rice husk fiber, ramie, kenaf, and jute fiber are utilized in many applications such as construction materials, the furniture sector, clothes, ropes, sewing thread, packaging materials, fishing nets, and paper production. Considerable endeavors have been undertaken to enhance the utilization

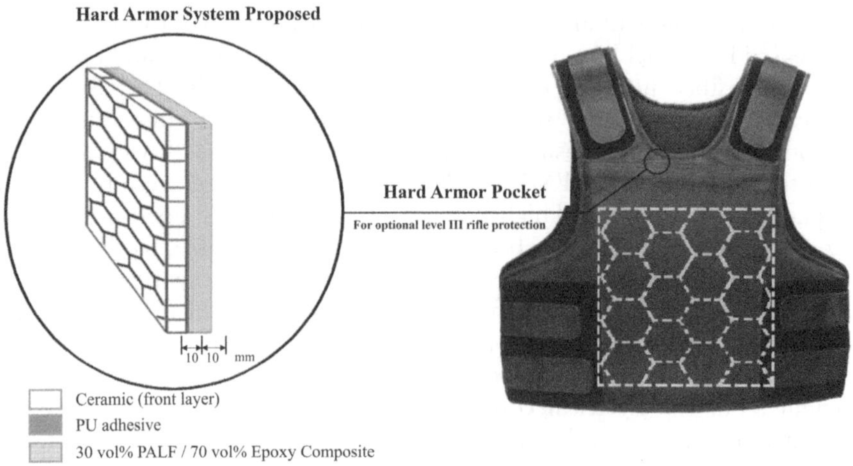

Figure 1.6 Schematic diagram of the hard armor system designed for ballistic protection [69].

of NF composites in the automotive sector, specifically in the domain of car interiors. In addition to its application in automobile interior parts, it is also utilized in the production of external auto body components [68].

Polyester composites reinforced with fique fibers were the subject of a comparable ballistic testing investigation by Luz et al. [69]. An evaluation was conducted on different volume fractions of fique fibers. The results showed that MAS works best with polyester composites with 30% fiber volume fractions and that these composites can stand in for Kevlar thanks to their superior impact resistance and energy absorption. This is an illustration of the hard armor system's structural layout (Figure 1.6), which is meant to protect against ballistic threats.

REFERENCES

[1] S. Hassanzadeh and H. Hasani. 2017. A Review on Milkweed Fiber Properties as a High-Potential Raw Material in Textile Applications. *Journal of Industrial Textiles*, 46(6), 1412–1436.

[2] R. Elayaraja, R. Sasikumar, M. Gokul, and T. Gowrisankar. 2020. Experimental Investigation of Prosopis Juliflora and Mango Tree with Epoxy Resin. *IOP Conference Series: Materials Science and Engineering*, 923(1), 012050.

[3] O. Faruk, A. K. Bledzki, H.-P. Fink, and M. Sain. 2012. Biocomposites Reinforced with Natural Fibers: 2000–2010. *Progress in Polymer Science*, 37(11), 1552–1596.

[4] M. Y. Khalid, A. Al Rashid, Z. U. Arif, W. Ahmed, H. Arshad, and A. A. Zaidi. 2021. Natural Fiber Reinforced Composites: Sustainable Materials for Emerging Applications. *Results in Engineering*, 11, 100263.

[5] G. Rajamurugan, R. Athithya, P. Krishnasamy, A. Nadeem, L. Kavinkumar, and R. Elayaraja 2024. Mechanical and Tribological Performance of Ramie Composite Reinforced with Perforated Al/Cu Foils. *Engineering Research Express*, 6(3), 032501.

[6] C. Q. J. Singh and G. Rajamurugan. 2021. Effect of Resin on Mechanical and Wear Performance of Wire Mesh-Reinforced Hydrophilic Fiber Composite. *Transactions of the Indian Institute of Metals*, 74(11), 2853–2867.

[7] P. Senthamaraikannan, S. S. Saravanakumar, V. P. Arthanarieswaran, and P. Sugumaran. 2016. Physico-Chemical Properties of New Cellulosic Fibers from the Bark of *Acacia planifrons*. *International Journal of Polymer Analysis and Characterization*, 21(3), 207–213.

[8] R. Sepe, F. Bollino, L. Boccarusso, and F. Caputo. 2018. Influence of Chemical Treatments on Mechanical Properties of Hemp Fiber Reinforced Composites. *Composite Part B: Engineering*, 133, 210–217.

[9] M. Z. Rong, M. Q. Zhang, Y. Liu, G. C. Yang, and H. M. Zeng. 2001. The Effect of Fiber Treatment on the Mechanical Properties of Unidirectional Sisal-Reinforced Epoxy Composites. *Composite Science and Technology*, 61(10), 1437–1447.

[10] K. C. Manikandan Nair, S. Thomas, and G. Groeninckx. 2001. Thermal and Dynamic Mechanical Analysis of Polystyrene Composites Reinforced with Short Sisal Fibres. *Composite Science and Technology*, 61(16), 2519–2529.

[11] M. A. Khan, M. M. Hassan, R. Taslima, and A. I. Mustafa. 2006. Role of Pretreatment with Potassium Permanganate and Urea on Mechanical and Degradable Properties of Photocured Coir (Cocos nucifera) fiber with 1,6-hexanediol diacrylate. *Journal of Applied Polymer Science*, 100(6), 4361–4368.

[12] H. U. Zaman, M. A. Khan, R. A. Khan, M. Arifur Rahman, L. R. Das, and M. Al-Mamun. 2010. Role of Potassium Permanganate and Urea on the Improvement of the Mechanical Properties of Jute Polypropylene Composites. *Fibers and Polymers*, 11(3), 455–463.

[13] V. S. Sreenivasan, D. Ravindran, V. Manikandan, and R. Narayanasamy. 2012. Influence of Fibre Treatments on Mechanical Properties of Short Sansevieria cylindrical/Polyester Composites. *Materials & Design*, 37, 111–121.

[14] R. Malkapuram, V. Kumar, and Y. S. Negi. 2009. Recent Development in Natural Fiber Reinforced Polypropylene Composites. *Journal of Reinforced Plastics and Composites*, 28(10), 1169–1189.

[15] S. Hattalli, A. Benaboura, F. Ham-Pichavant, A. Nourmamode, and A. Castellan. 2002. Adding Value to Alfa Grass (*Stipa tenacissima* L.) Soda Lignin as Phenolic Resins 1. Lignin Characterization. *Polymer Degradation and Stability*, 76(2), 259–264.

[16] M. Jawaid and H. P. S. Abdul Khalil. 2011. Cellulosic/Synthetic Fibre Reinforced Polymer Hybrid Composites: A Review. *Carbohydrate Polymers*, 86(1), 1–18.

[17] A. K. Bhunia, D. Mondal, K. R. Sahu, and A. K. Mondal. 2023. Characterization of New Natural Cellulosic Fibers from *Cyperus compactus* Retz. (Cyperaceae) Plant. *Carbohydrate Polymer Technologies and Applications*, 5, 100286. https://doi.org/10.1016/j.carpta.2023.100286

[18] K. Mayandi, N. Rajini, P. Pitchipoo, J. T. W. Jappes, and A. V. Rajulu. 2016. Extraction and Characterization of New Natural Lignocellulosic Fiber *Cyperus pangorei*. *International Journal of Polymer Analysis and Characterization*, 21(2), 175–183.

[19] R. M. Rowell. 2014. The Use of Biomass to Produce Bio-Based Composites and Building Materials. *Advances in Bio Refineries, Elsevier*, 803–818. https://doi.org/10.1533/9780857097385.2.803
[20] M. Maran, R. Kumar, P. Senthamaraikannan, S. S. Saravanakumar, S. Nagarajan, M. R. Sanjay, and S. Siengchin. 2022. Suitability Evaluation of *Sida mysorensis* Plant Fiber as Reinforcement in Polymer Composite. *Journal of Natural Fibers*, 19(5), 1659–1669.
[21] A. Khan, R. Vijay, D. Lenin Singaravelu, M. R. Sanjay, S. Siengchin, M. Jawaid, K. A. Alamry, A. M. Asiri. 2022. Extraction and Characterization of Natural Fibers from *Citrullus lanatus* Climber. *Journal of Natural Fibers*, 19(2), 621–629.
[22] E. Sarikaya, H. Çallioğlu, and H. Demirel. 2019. Production of Epoxy Composites Reinforced by Different Natural Fibers and their Mechanical Properties. *Composite Part B: Engineering*, 167, 461–466.
[23] A. Khan, R. Vijay, D. Lenin Singaravelu, M. R. Sanjay, S. Siengchin, F. Verpoort, K. A. Alamry, and A. M. Asiri. 2021. Characterization of Natural Fibers from *Cortaderia selloana* Grass (Pampas) as Reinforcement Material for the Production of the Composites. *Journal of Natural Fibers*, 18(11), 1893–1901.
[24] V. S. Sreenivasan, S. Somasundaram, D. Ravindran, V. Manikandan, and R. Narayanasamy. 2011. Microstructural, Physico-Chemical and Mechanical Characterisation of *Sansevieria cylindrica* Fibres: An Exploratory Investigation. *Materials & Design*, 32(1), 453–461.
[25] A. K. Mohanty, M. Misra, and L. T. Drzal. 2001. Surface Modifications of Natural Fibers and Performance of the Resulting Biocomposites: An Overview. *Composite Interfaces*, 8(5), 313–343.
[26] F. Martí-Ferrer, F. Vilaplana, A. Ribes-Greus, A. Benedito-Borrás, and C. Sanz-Box. 2006. Flour Rice Husk as Filler in Block Copolymer Polypropylene: Effect of Different Coupling Agents. *Journal of Applied Polymer Science*, 99(4), 41823–41831.
[27] W. Fatriasari and E. Hermiati. 2008. Analysis of Fiber Morphology and Physical-Chemical Properties of Six Species of Bamboo as Raw Material for Pulp and Paper. *Journal of Forestry Science and Technology*, 1(2), 67–72.
[28] N. Sumrith, L. Techawinyutham, M. R. Sanjay, R. Dangtungee, and S. Siengchin. 2020. Characterization of Alkaline and Silane Treated Fibers of 'Water Hyacinth Plants' and Reinforcement of 'Water Hyacinth Fibers' with Bioepoxy to Develop Fully Biobased Sustainable Eco friendly Composites. *Journal of Polymers and the Environment*, 28(10), 2749–2760.
[29] P. Narayanasamy et al. 2020. Characterization of Vetiver Grass Fibers and their Potential Exploration in Bio Polymer Composite Fabrication. *Journal of Polymers and the Environment*, 28(8), 2061–2071.
[30] P. Madhu, M. R. Sanjay, M. Jawaid, S. Siengchin, A. Khan, and C. I. Pruncu. 2020. A New Study on Effect of Various Chemical Treatments on Agave Americana Fiber for Composite Reinforcement: Physico-Chemical, Thermal, Mechanical and Morphological Properties. *Polymer Testing*, 85, 106437.
[31] G. Silva, S. Kim, R. Aguilar, and J. Nakamatsu. 2020. Natural Fibers as Reinforcement Additives for Geopolymers – A Review of Potential Eco-Friendly Applications to the Construction Industry. *Sustainable Materials and Technologies*, 23, e00132.

[32] M. Rajabi, A. Ali, M. McConnell, and J. Cabral. 2020. Keratinous Materials: Structures and Functions in Biomedical Applications. *Materials Science and Engineering: C*, 110, 110612.

[33] N. Venkateshwaran and A. ElayaPerumal. 2012. Mechanical and Water Absorption Properties of Woven Jute/Banana Hybrid Composites. *Fibers and Polymers*, 13(7), 907–914.

[34] Q. Mo, Y. Huang, L. Ma, W. Lai, Y. Zheng, Y. Li, M. Xu, Z. Huang. 2022. Study on Microwave Curing of Unsaturated Polyester Resin and Its Composites Containing Calcium Carbonate. *Polymers*, 14(13), 2598.

[35] B. Wielage, T. Lampke, H. Utschick, and F. Soergel. 2003. Processing of Natural-Fibre Reinforced Polymers and the Resulting Dynamic–Mechanical Properties. *Journal of Materials Processing Technology*, 139(1–3), 140–146.

[36] A. Paul, K. Joseph, and S. Thomas. 1997. Effect of surface Treatments on the Electrical Properties of Low-Density Polyethylene Composites Reinforced with Short Sisal Fibers. *Composites Science and Technology*, 57(1), 67–79.

[37] D. Rouison, M. Sain, and M. Couturier. 2004. Resin Transfer Molding of Natural Fiber Reinforced Composites: Cure Simulation. *Composites Science and Technology*, 64(5), 629–644.

[38] F. M. B. Coutinho and T. H. S. Costa. 1999. Performance of Polypropylene–Wood Fiber Composites. *Polymer Testing*, 18(8), 581–587.

[39] N. F. Zaaba and H. Ismail. 2019. Thermoplastic/Natural Filler Composites: A Short Review. *Journal of Physical Science*, 30(Suppl. 1), 81–99.

[40] A. Valadez-Gonzalez, J. M. Cervantes-Uc, R. Olayo, and P. J. Herrera-Franco. 1999. Effect of Fiber Surface Treatment on the Fiber–Matrix Bond Strength of Natural Fiber Reinforced Composites. *Composites Part B: Engineering*, 30(3), 309–320.

[41] M. K. Huda and I. Widiastuti. 2021. Natural Fiber Reinforced Polymer in Automotive Application: A Systematic Literature Review. *Journal of Physics: Conference Series*, 1808(1), 012015.

[42] H. Bisaria, M. K. Gupta, P. Shandilya, and R. K. Srivastava. 2015. Effect of Fibre Length on Mechanical Properties of Randomly Oriented Short Jute Fibre Reinforced Epoxy Composite. *Materials Today: Proceedings*, 2(4–5), 1193–1199.

[43] A. Kumre, R. S. Rana, and R. Purohit. 2017. A Review on Mechanical Property of Sisal Glass Fiber Reinforced Polymer Composites. *Materials Today: Proceedings*, 4(2), 3466–3476.

[44] A. C. N. Singleton, C. A. Baillie, P. W. R. Beaumont, and T. Peijs. 2003. On the Mechanical Properties, Deformation and Fracture of a Natural Fibre/Recycled Polymer Composite. *Composites Part B: Engineering*, 34(6), 519–526.

[45] Q. Liu and M. Hughes. 2008. The Fracture Behaviour and Toughness of Woven Flax Fibre Reinforced Epoxy Composites. *Composites Part A: Applied Science and Manufacturing*, 39(10), 1644–1652.

[46] M. A. Sawpan, K. L. Pickering, and A. Fernyhough. 2011. Effect of Various Chemical Treatments on the Fibre Structure and Tensile Properties of Industrial Hemp Fibres. *Composites Part A: Applied Science and Manufacturing*, 42(8), 888–895.

[47] P. Saha, S. Manna, S. R. Chowdhury, R. Sen, D. Roy, and B. Adhikari. 2010. Enhancement of Tensile Strength of Lignocellulosic Jute Fibers by Alkali-Steam Treatment. *Bioresource Technology*, 101(9), 3182–3187.

[48] K. Obi Reddy, C. Uma Maheswari, M. Shukla, J. I. Song, and A. Varada Rajulu. 2013. Tensile and Structural Characterization of Alkali Treated Borassus Fruit Fine Fibers. *Composites Part B: Engineering*, 44(1), 433–438.

[49] Y. Seki, M. Sarikanat, K. Sever, and C. Durmuşkahya. 2013. Extraction and Properties of *Ferula communis* (chakshir) Fibers as Novel Reinforcement for Composite Materials. *Composites Part B: Engineering*, 44(1), 517–523.

[50] K. S. Kumar, I. Siva, N. Rajini, P. Jeyaraj, and J. W. Jappes. 2014. Tensile, Impact, and Vibration Properties of Coconut Sheath/Sisal Hybrid Composites: Effect of Stacking Sequence. *Journal of Reinforced Plastics and Composites*, 33(19), 1802–1812.

[51] C. Bennet, N. Rajini, J. W. Jappes, I. Siva, V. Sreenivasan, and S. Amico. 2015. Effect of the Stacking Sequence on Vibrational Behavior of Sansevieria Cylindrica/Coconut Sheath Polyester Hybrid Composites. *Journal of Reinforced Plastics and Composites*, 34(4), 293–306.

[52] K. Senthil Kumar, I. Siva, P. Jeyaraj, J. T. Winowlin Jappes, S. C. Amico, and N. Rajini. 2014. Synergy of Fiber Length and Content on Free Vibration and Damping Behavior of Natural Fiber Reinforced Polyester Composite Beams. *Materials & Design*, 56, 379–386.

[53] G. Kalusuraman, S. Irullappasamy, and M. Aslan. 2019. Free Vibration and Damping Behaviour of Surface Treated Luffa/USP Composite Beams. *SSRN Electronic Journal*. Available at SSRN: https://ssrn.com/abstract=3654005 or https://doi.org/10.2139/ssrn.3654005

[54] Y. S. Munde, R. B. Ingle, and I. Siva. 2018. Investigation to Appraise the Vibration and Damping Characteristics of Coir Fibre Reinforced Polypropylene Composites. *Advances in Materials and Processing Technologies*, 4(4), 639–650.

[55] M. Rajesh and J. Pitchaimani. 2017. Experimental Investigation on Buckling and Free Vibration Behavior of Woven Natural Fiber Fabric Composite Under Axial Compression. *Composite Structures*, 163, 302–311.

[56] K. Senthil Kumar, I. Siva, N. Rajini, J. T. Winowlin Jappes, and S. C. Amico. 2016. Layering Pattern Effects on Vibrational Behavior of Coconut Sheath/Banana Fiber Hybrid Composites. *Materials & Design*, 90, 795–803.

[57] A. Etaati, S. A. Mehdizadeh, H. Wang, and S. Pather. 2014. Vibration Damping Characteristics of Short Hemp Fibre Thermoplastic Composites. *Journal of Reinforced Plastics and Composites*, 33(4), 330–341.

[58] P. C. Jena. 2018. Free Vibration Analysis of Short Bamboo Fiber Based Polymer Composite Beam Structure. *Materials Today: Proceedings*, 5(2), 5870–5875.

[59] S. S. Rana, M. K. Gupta, and R. K. Srivastava. 2017. Effect of Variation in Frequencies on Dynamic Mechanical Properties of Short Sisal Fibre Reinforced Epoxy Composite. *Materials Today: Proceedings*, 4(2), 3387–3396.

[60] S. Prabhakaran, V. Krishnaraj, M. S. Kumar, and R. Zitoune. 2014. Sound and Vibration Damping Properties of Flax Fiber Reinforced Composites. *Procedia Engineering*, 97, 573–581.

[61] Y. S. Munde, R. B. Ingle, and I. Siva. 2019. Vibration Damping and Acoustic Characteristics of Sisal Fibre–Reinforced Polypropylene Composite. *Noise & Vibration Worldwide*, 50(1), 13–21.

[62] G. S. Divya, A. Kakhandaki, and B. Suresha. 2014. Wear Behavior of Coir Reinforced Treated and Untreated Hybrid Composites. Available: www.ijird.com

[63] B. F. Yousif. 2009. Frictional and Wear Performance of Polyester Composites Based on Coir Fibres. *Proceedings of the Institution of Mechanical Engineers, Part J: Journal of Engineering Tribology*, 223(1), 51–59.
[64] S. S. Todkar and S. A. Patil. 2019. Review on Mechanical Properties Evaluation of Pineapple Leaf Fibre (PALF) Reinforced Polymer Composites. *Composites Part B: Engineering*, 174, 106927.
[65] M. R. Sanjay, G. R. Arpitha, L. L. Naik, K. Gopalakrishna, and B. Yogesha. 2016. Applications of Natural Fibers and Its Composites: An Overview. *Natural Resources*, 7(3), 108–114.
[66] J. Holbery and D. Houston. 2006. Natural-Fiber-Reinforced Polymer Composites in Automotive Applications. *JOM*, 58(11), 80–86.
[67] L. Mohammed, M. N. M. Ansari, G. Pua, M. Jawaid, and M. S. Islam. 2015. A Review on Natural Fiber Reinforced Polymer Composite and Its Applications. *International Journal of Polymer Science*, 2015, 1–15.
[68] A. C. Pereira, F. S. Assis, F.D.C. Garcia Filho, M.S. Oliveira, E.S. Lima, H.A.C. Lopera, and S.N. Monteiro 2019. Evaluation of the Projectile's Loss of Energy in Polyester Composite Reinforced with Fique Fiber and Fabric. *Materials Research*, 22(suppl 1). https://doi.org/10.1590/1980-5373-MR-2019-0146
[69] F. Luz, F. Garcia Filho, M. Oliveira, L. Nascimento, and S. Monteiro. 2020. Composites with Natural Fibers and Conventional Materials Applied in a Hard Armor: A Comparison. *Polymers*, 12(9), 1920.

Chapter 2

Optimization of carbon sphere-based polymer composite coating for solar collector using response surface methodology

Radhika Panickar, Albin Joseph, Shijo Thomas, C. B. Sobhan, and Sivaji Chakravorti

2.1 INTRODUCTION

The drastic increment in the emission of carbon has become a severe environmental issue. As per the reports of the Intergovernmental Panel for Climate Change, the global surface temperature has increased by nearly 1.5°C since 1850. Even though this value looks very less, its impact is very high. Experts attribute the global climatic change to the surplus emission of carbon into the atmosphere. It is well known that fossil fuels are the main source of energy and a major source of carbon emission. Considering the ecological aspects, the usage of fossil fuels needs to be limited. One of the potential strategies to reduce carbon emissions is by promoting renewable energy sources. Among the various renewable energy sources, solar energy is a widely acknowledged candidate for the same. Photovoltaic panels are one of the commonly used techniques for the harvesting of solar energy. However, the efficiency of the PV system is limited to 20% since it utilizes the visible rays in the solar spectrum alone. Solar thermal conversion is yet another harvesting technique for converting solar energy to useful thermal energy. The solar thermal conversion can be utilized in association with various applications, such as thermal energy storage [1], solar water heaters [2], solar-powered sterilization of medical instruments [3], steam generation [4], etc., to name a few. Selective absorber coatings have significant importance in improving the efficiency of solar thermal collectors for photothermal conversion [5–7] by enhancing the spectral absorptivity and increasing the range of incident wavelengths that are absorbed.

A traditional solar collector where the solar energy is converted to thermal energy by employing a solar selective coating on the absorber surface of the solar collector is termed the surface absorption technique. The thermal energy from the absorber is then transferred to the heat transfer fluid by conduction/convection/radiation and is used for various applications. As the thermal conversion of solar radiation is achieved by the coating employed, its optical properties have high importance. Different solar selective materials with high solar absorption capabilities have been reported in recent years. Selective solar absorber coatings based on metals, metal oxides,

 DOI: 10.1201/9781003527275-2

cermet, and composite films prepared by various means with high transmittivity and low reflectivity in visible and near IR regions have been reported extensively [8, 9]. CrN_xO_y [8], Ni–Al_2O_3 [10] HfO_x/Mo/HfO_2 [11], TiAlSiN/TiAlSiON/SiO_2 [12], and composite coatings of Cr–Cr_2O_3 and Mo–Al_2O_3 [13] are a few examples. In the surface absorption collector, the thermal energy is transferred to the heat transfer fluid after conduction through the absorber tube (especially in the case of a parabolic collector, evacuated tube collector, dish collector, etc.) and convection from the absorber tube. However, the conduction and convection of thermal energy to the ambient, due to a rise in absorber surface temperature, result in energy losses and reduce the overall efficiency of the system. To eliminate this conduction and convection loss, the concept of direct absorption of solar radiation by the working fluid was introduced and gained wide attention. In a direct absorption solar collector, the heat transfer fluid is functionalized to absorb the solar radiation and convert it to thermal energy.

The direct absorption of solar radiation by the heat transfer fluid reduces the intermediate heat transfer losses as in surface absorption, and hence the efficiency was found to be comparatively high. The functionalization of the working fluid to enhance its solar absorptivity is achieved by using nanofluid as the working fluid. The efficiency of the parabolic solar trough collectors is significantly influenced by the type of nanoparticle and its concentration in the working fluid. It has been reported that nanofluids with Au nanoparticles showed high photostability, and $NiFe_2O_4$ improved the thermal efficiency of the parabolic solar collector. The efficiency of the collector could be further increased by having coatings of higher solar absorptivity. Carbon-based materials have exhibited superior solar absorption. The present study attempts to improve the solar absorptivity of a hybrid absorption system using a CS-based coating on the absorber surface. The introduction of CS could enhance the absorptivity of the collector and subsequently the thermal performance [14].

CS have a wide range of applications [15–17]. In this chapter, a polymer-based CS/PMMA/epoxy coating was prepared and its influence on the hybrid absorption system was analyzed. A schematic of the proposed system is presented in Figure 2.1. The CS was coated on the interior face of the absorber tube, using PMMA as the binder. The CS for the absorber coating were synthesized by a non-catalytic chemical vapor deposition (CVD) technique. The preparation of the selective absorber coating of CS/PMMA/epoxy was carried out using the dip-coating technique. To optimize and study the interaction of various coating process parameters, a statistical approach based on Design of Experiments (DOE) was employed. The photothermal conversion efficiency of the coated substrate was experimentally investigated and compared with that of the uncoated substrate. The experiments were conducted in the southern part of India, where the average radiation during clear skies is 900 W/m^2.

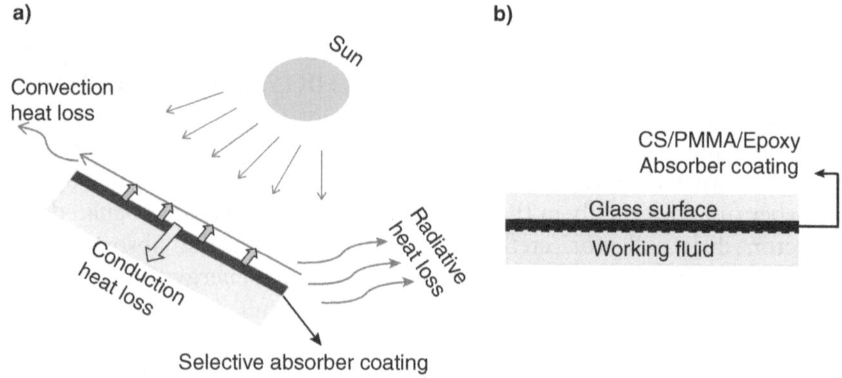

Figure 2.1 Schematic of (a) conventional absorber coating (b) CS/PMMA/epoxy coating used in the present work.

2.2 MATERIALS AND METHODS

2.2.1 Materials

The CS were synthesized using the non-catalytic CVD method. Hydrogen, acetylene, and argon of 99.9% purity were supplied by Bhoruka Gases Ltd., India. Polymethyl methacrylate (PMMA) was purchased from Himedia, India. Toluene for dissolving PMMA was purchased from Fisher Scientific, and epoxy resin was purchased from Hakson Pvt. Ltd., India.

2.2.2 Synthesis of carbon spheres

Figure 2.2 presents the schematic of a thermal CVD setup for CS synthesis. The synthesis of CS was carried out in a tubular furnace with a one-meter-long quartz tube. Acetylene as the carbon source, and argon and hydrogen as the carrier gases, were fed into the quartz tube for the synthesis of CS.

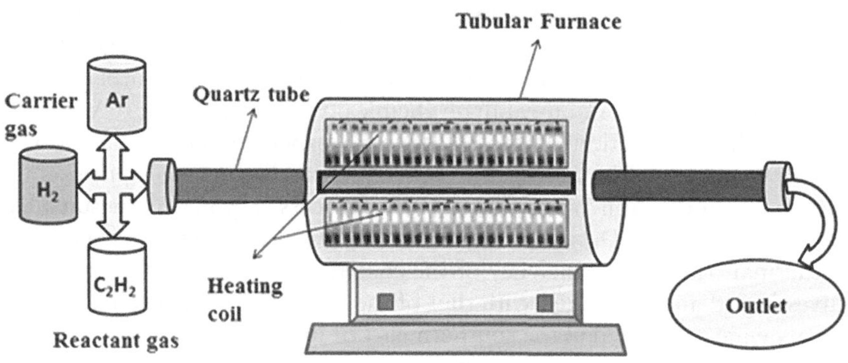

Figure 2.2 Schematic of thermal CVD setup for carbon sphere synthesis.

Hydrogen, acetylene, and argon gases were supplied at steady flow rates of 30, 100, and 200 sccm, respectively.

The synthesis process was carried out in three steps involving; (i) heating the chamber to 950°C under Ar and H_2 atmosphere, (ii) introducing acetylene in the chamber for 30 min, after attaining the required deposition temperature, and (iii) cooling the chamber under an inert atmosphere to room temperature. During the synthesis, carbon was deposited inside the walls of the quartz chamber and was collected for further characterization. A detailed synthesis of CS has been reported in our previous work [18]. Using this method, up to a few grams of relatively uniform-sized CS can be synthesized within 30 min of carbon precursor flow.

2.2.3 Preparation of CS/PMMA/epoxy thin film

The synthesized CS were dispersed in PMMA, and the CS/PMMA/epoxy absorber coating was prepared using the dip-coating technique. Dip coating is one of the simplest and commercially adopted methods for the synthesis of uniform thin films [19, 20]. The dip-coating technique involves controlled dipping and the subsequent withdrawal of the substrate from the precursor. In the present study, the composite coating was achieved using a dip-coating apparatus (Prompt-DCU200). Figure 2.3 gives the schematic representation

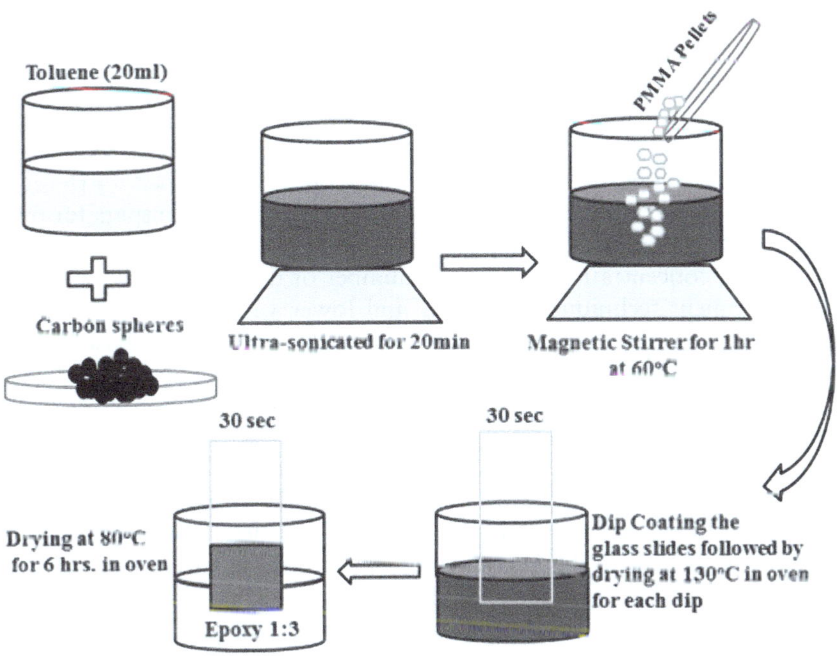

Figure 2.3 Schematic representation of CS/PMMA/epoxy composite preparation.

of the steps through which the dip coating of CS/PMMA/epoxy thin film on glass substrate was performed. The CS/PMMA films were prepared according to the proposed design matrix values for different variables given in Table 2.2. A total of 15 CS/PMMA films were prepared on glass slides with the concentration of CS, PMMA, and the number of dips mentioned in the design matrix. Each sample is prepared by adding a prescribed amount of CS and PMMA pellets to 20 mL of toluene from Table 2.2. Toluene is the solvent for PMMA that is heated at 60°C with continuous stirring for 1 hr to dissolve the PMMA pellets. The solution was cooled to room temperature, and dip coating was performed on cleaned glass substrates. The glass slides were cleaned in an ultrasonic bath for about 15 min to remove impurities and non-homogeneities on the surface. During dip coating, each dip was done for 30 s, followed by 10 min drying in the open air and 15 min drying in a hot air oven at 130°C. CS/PMMA coated glass slides was then dip-coated in epoxy and dried at 80°C in the oven for 6 hrs to obtain CS/PMMA/epoxy thin films on glass substrates.

It is evident from the literature that various parameters, concentration of the precursor, dipping parameters, etc., influence the performance of the coating [21]. Hence, the investigation of the variation of coating process parameters and their interaction has high importance. To study the interaction of coating parameters, a statistical approach employing the DOE was adopted. In addition to the investigations on the interaction of coating parameters, the tool enables optimization of the various coating parameters to maximize the absorptivity of the absorber.

DOE is a statistical procedure designed to plan an experiment with many input variables effectively. In this study, among various tools included in DOE, response surface methodology (RSM)-based statistical optimization is used to study the correlation between the three influential process variables. The Central Composite Design (CCD) based on RSM was used for the evaluation of the process variables to obtain the experimental outputs for optimization. The identified process variables for the coating were concentration of PMMA, concentration of CS, and number of dips. As per the design of the experiment technique, the upper and lower value of each variable is fixed by conducting a literature survey and experimental trials. Based on the upper and lower values of each variable, a design matrix containing a set of experimental runs will be generated. Table 2.1 presents the upper and lower values and intermediate values of identified coating variables. The design matrix generated is presented in Table 2.2. CCD enables us to estimate the curvature effect by considering a group of axial/star points as shown in Figure 2.4. In the present work, CCD with three independent variables was applied to investigate the effect of influential process variables to achieve maximum absorbance for the CS/PMMA coating. In this case, CCD involves 2^N factorial points, 2N axial points, and 1 center point, where N is the number of process variables.

Table 2.1 Range and level of process variables for DOE

		Levels				
Sl No	*Process parameter*	*−1.6871 (−α)*	*−1*	*0*	*+1*	*+1.6871 (+α)*
1	Concentration of PMMA (mg)	2500	3209.4	4250	5290.6	6000
2	Concentration of CS (mg)	50	80.4	125	169.6	200
3	Number of dip	1	2	3	4	5

Table 2.2 Design matrix from DOE and experimental response outputs

Std:	*Run*	*Factor 1*	*Factor 2*	*Factor 3*	*Response*
Sample No:	*Run*	*PMMA (mg)*	*Carbon spheres (mg)*	*No. of dip*	*SWAF*
1	7	2500	125	3	40
2	10	4250	125	1	42
3	18	4250	125	5	55
4	6	4250	125	3	44
5	1	4250	200	3	48
6	2	3209.4	80.4	2	34
7	16	3209.4	80.4	4	50
8	11	3209.4	169.6	2	62
9	20	3209.4	169.6	4	71
10	9	5290.6	80.4	4	84
11	14	5290.6	80.4	2	69
12	0	5290.6	169.6	2	77
13	19	5290.6	169.6	4	93
14	12	4250	50	3	48
15	13	6000	125	3	94
4.a	3	4250	125	3	41
4.b	4	4250	125	3	46
1.c	17	4250	125	3	45
4.d	15	4250	125	3	48
4.e	5	4250	125	3	43

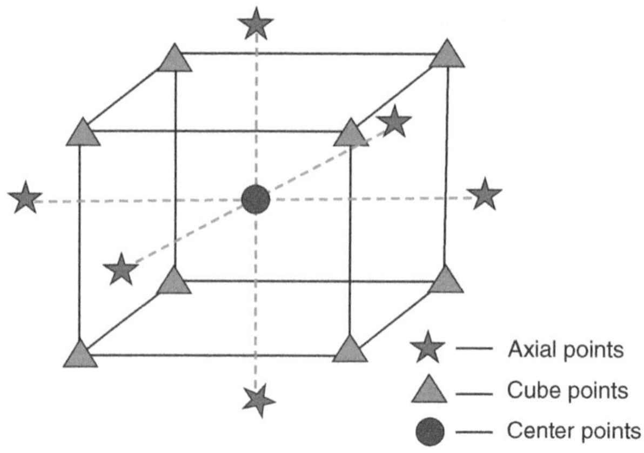

Figure 2.4 Design of CCD.

2.2.4 Solar thermal conversion experimentation

Solar thermal conversion performance of the coatings was examined using a lab-made setup, as shown in Figure 2.5(a). Once the solar weighted absorption fraction (SWAF) of samples was quantified based on the design matrix, critical samples were selected. The critical samples were selected as per the SWAF observed. Two samples each that exhibited maximum SWAF, minimum SWAF, and intermediate SWAF were considered as the six critical samples.

The six critical samples were assigned sample numbers as discussed later in the chapter. The coating was achieved on the inner surface of the glass tube of dimensions 15 cm in length and 1.8 cm in diameter. As shown in Figure 2.5(b), the test samples coated with CS/PMMA/epoxy were arranged in a wooden stand. DI water was employed as the working fluid in the samples. The coated samples were then exposed to solar radiation, and the temperature rise was noted using a calibrated T-type thermocouple.

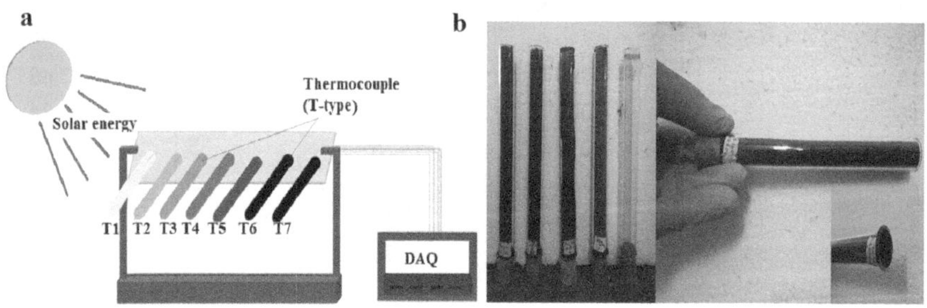

Figure 2.5 (a) Experimental setup for photothermal studies (b) CS/PMMA/epoxy coating.

The results obtained for the coated samples were then compared with those for an uncoated glass tube sample with DI water.

2.3 RESULTS AND DISCUSSIONS

2.3.1 Characterization of CS

Scanning Electron Microscopy (SEM) (Hitachi SU6600–FESEM) was used to study the morphology and structure of the synthesized CS. The CS is an amorphous carbon material with a spheroidal shape. Figure 2.6(a) is the SEM image of the CS synthesized at 950°C that shows a smooth texture and spheroidal morphology. The relative size of the CS synthesized was measured, and it ranges from 200 to 350 nm. Further morphology analysis carried out using a Transmission Electron Microscope (TEM) (JEM 2100, JEOL Ltd.) is given in Figure 2.6(b), which provides more clarity on the morphology of the CS synthesized. The CS obtained is conglomerated to form chain-like structures due to the fast growth of carbon nuclei under the deposition temperature [20].

The phase identification of the CS was carried out using an X-ray diffractometer (Rigaku Miniflex 600) using Cu-kα radiation of wavelength $\lambda = 1.5406$ Å. Figure 2.7(a) shows the XRD spectrum of the CS having broad peaks with 2θ values at 24.65° and 43.07°, which represent the graphitic planes of (002) and (100), respectively. The broad peaks from the XRD spectra confirm that the CS synthesized is amorphous carbon material.

The vibrational modes and chemical composition of the synthesized CS are characterized using Raman spectroscopy. The Raman spectrum was obtained from WITEC Alpha 300 RA with a maximum power of 70 mW and a DPSS laser of 532 nm. Using Raman spectroscopy, the graphitic degree (I_D/I_G) of the CS synthesized was measured as 0.85. Figure 2.7(b) gives the Raman spectrum showing two broad peaks, a strong peak at 1578 cm^{-1} and

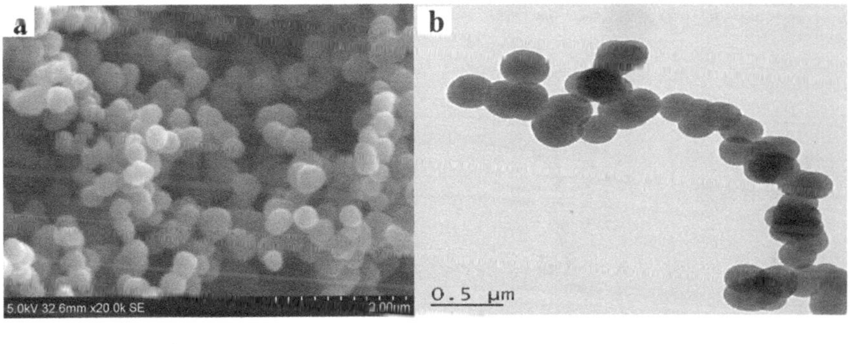

Figure 2.6 Morphology study of carbon spheres using (a) Scanning Electron Microscopy (SEM) and (b) Transmission Electron Microscopy (TEM).

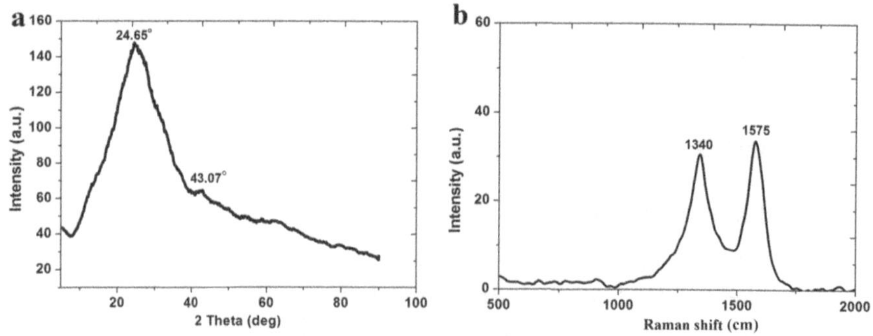

Figure 2.7 (a) X-Ray powder diffraction pattern [18] and (b) Raman Spectrum of carbon spheres.

a weaker peak at 1336 cm^{-1}. The high graphitization degree and broad peaks in the Raman spectrum also confirm that the sample is amorphous carbon material.

2.3.2 Optical characterization of CS/PMMA/epoxy films

The optical properties of the coated glass slides were examined using a UV–Vis-NIR spectrometer (Shimadzu, UV-2401 PC). Figure 2.8 presents the spectrum of absorbance (α), reflectance(r), and transmittance (τ). From the

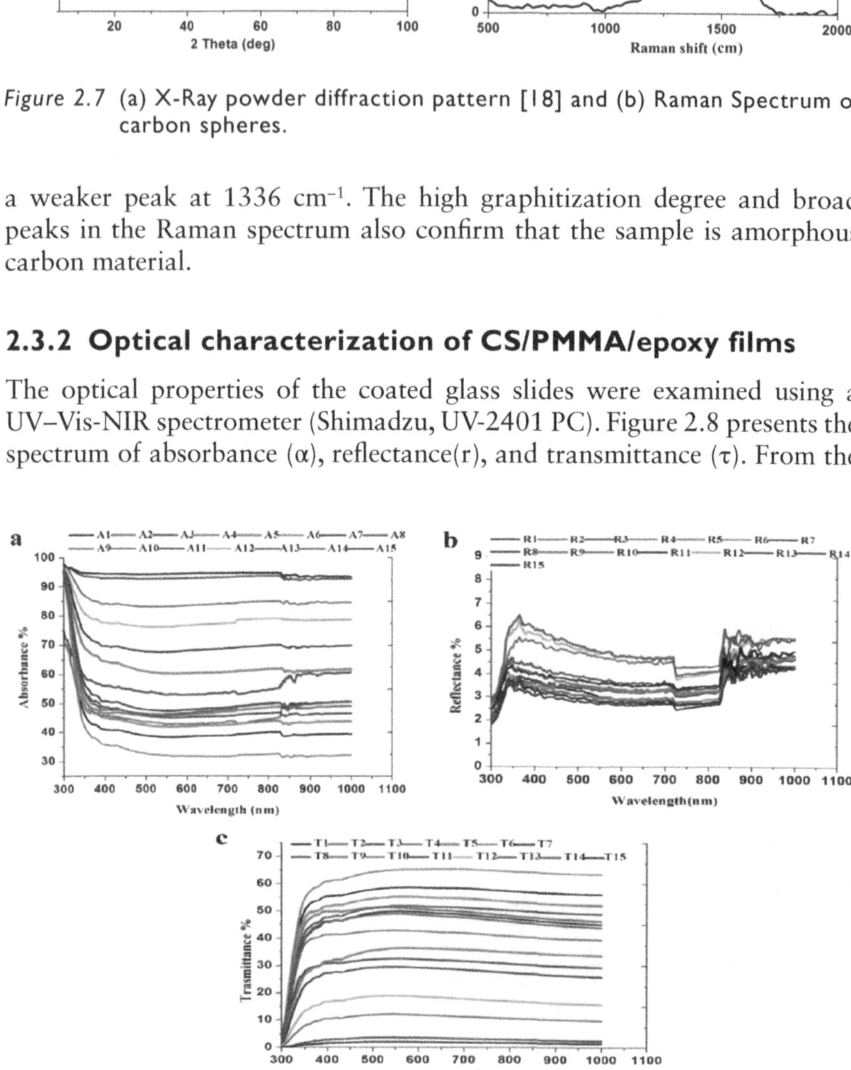

Figure 2.8 Optical properties of glass samples obtained from UV-visible spectroscopy (a) absorbance, (b) reflectance, and (c) transmittance.

spectral properties of the samples, the solar transmittance (τ_o), reflectance (r_o), and absorbance (α_o) are calculated using the equations (2.1)–(2.3). In addition, the SWAF was obtained from equation (2.3). SWAF estimated for each sample in the design matrix is presented in Table 2.2.

$$\tau_o = \frac{\sum_{\lambda l=300}^{\lambda h=1100} S_\lambda \tau(\lambda)\Delta\lambda}{\sum_{\lambda l=300}^{\lambda h=1100} S_\lambda \Delta\lambda} \quad (2.1)$$

$$r_o = \frac{\sum_{\lambda l=300}^{\lambda h=1100} S_\lambda r(\lambda)\Delta\lambda}{\sum_{\lambda l=300}^{\lambda h=1100} S_\lambda \Delta\lambda} \quad (2.2)$$

$$\alpha_o = \left[1 - \left(\tau_0 + r_0\right)\right] \quad (2.3)$$

where S_λ is the relative spectral density of the solar radiation, $\tau(\lambda)$ is the measured spectral transmittance of the glass, $r(\lambda)$ is the measured spectral reflectance of the glass, and $\Delta\lambda$ is the wavelength interval.

From the response obtained from the optical property studies of different samples, a maximum absorbance of 94% was obtained for sample 15 with the concentration of CS and PMMA as 125 and 6,000 mg, respectively, and 3 dips for achieving the coating. The minimum absorbance was obtained for sample 6 with a SWAF of 34% with a concentration of CS and PMMA as 80.4 and 3209.4 mg and the number of dips = 2. All 20 experimental response outputs obtained from the optical characterization are presented in Table 2.2.

2.3.3 Evaluation using ANOVA

The DOE tool enables investigation of the interaction of each process variable on a part by developing a regression model from the design matrix and the corresponding output. The Analysis of Variance (ANOVA) evaluates the significance of each process variable and the model. The ANOVA results are summarized in Table 2.3. The results are expressed using various statistical quantities such as the *p*-value (probability), F- value, and degrees of freedom (df). Using ANOVA, the mean square values are calculated by dividing the sum of squares of each variation source by their degrees of freedom, and the significance level (α value = 0.05) was used to test the significance of process variables. F value gives the significance of each process parameter and the regression model developed. As can be seen from the ANOVA table, the concentration of PMMA exhibits a higher value of F and hence is the

Table 2.3 Analysis of variance table

Source	*Sum of squares*	*Df*	*Mean square*	*F value*	*p-value Prob > F*	
Model	9253.36	9	1028.15	73.93	<0.0001	Significant
A-PMMA	3939.73	1	3939.73	283.30	<0.0001	
B-CS	1691.32	1	1691.32	121.62	<0.0001	
C-dip	2368.69	1	2368.69	170.33	<0.0001	
AB	231.12	1	231.12	16.62	0.0022	
AC	105.13	1	105.13	7.56	0.0205	
BC	14.04	1	14.04	1.01	0.3386	
A^2	71.21	1	71.21	5.12	0.0471	
B^2	554.04	1	554.04	39.84	<0.0001	
C^2	415.50	1	415.50	29.88	0.0003	
Residual	139.07	10	13.91			
Lack of fit	115.73	5	23.15	4.96	0.0518	Not significant
Pure error	23.33	5	4.67	R^2	0.9852	
Std. dev	3.73			Adj R^2	0.9719	
CV %	6.10			Pred R^2	0.8935	

most significant parameter on output response. It is interesting to note that the least significant parameter is the concentration of CS. The reason behind this may be the superior binding effect of PMMA to keep the CS particle on the thin film. Moreover, the number of dips also significantly influences the absorptivity of the coating.

As far as the significance of the interaction effect is considered, it could be noted that the interaction of concentration of PMMA and CS is the highest with an F value of 16.62. As mentioned, a regression equation was developed using the design matrix and the output response; the significance of the same could also be analyzed from the ANOVA. The F value calculated for the model is 73.93, and a very small probability value (p-value) of <0.0001 shows the very high significance of the model. In the model, the F value is used to compare the curvature variance with the residual variance and the probability > F (p-value).

The goodness of fit for the model can also be determined from the correlation coefficient (R^2), which gives a value of 0.9852. Figure 2.9 gives a comparison of the experimental data with the predicted data, which shows good agreement between the model and experimental data. Therefore, the statistical method gives a good correlation between the experimental and predicted results. The developed empirical model using the RSM is presented in equation (2.4).

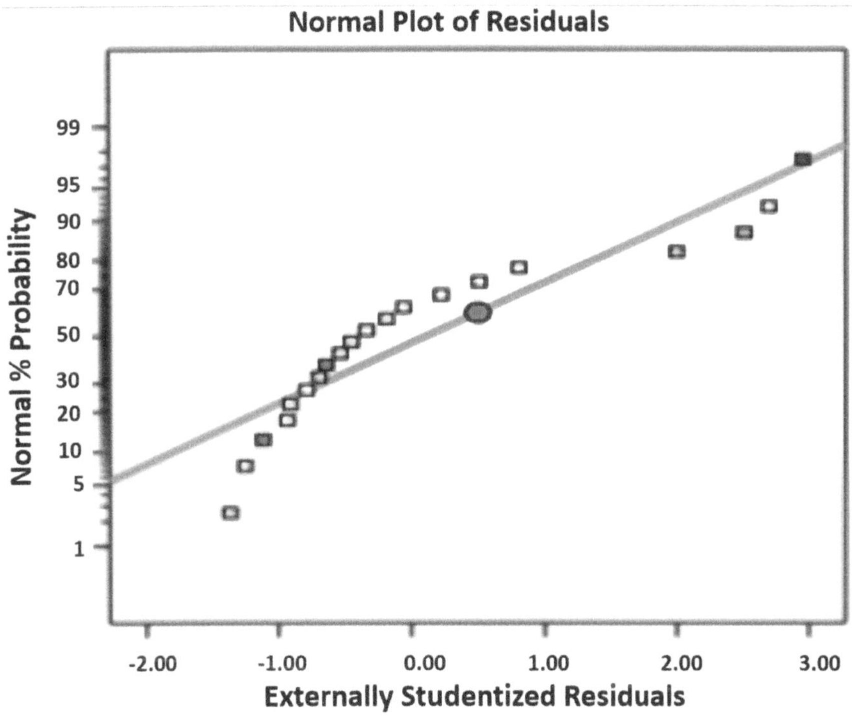

Figure 2.9 The plot of predicted vs actual data.

$$\begin{aligned}\text{SWAF} = {} & -272.67443 + 0.057040 \times \text{PMMA} + 1.44631 \\ & \times \text{CS} + 43.18245 \times \text{Dip} - 0.000116 \times \text{PMMA} \\ & \times \text{CS} - 0.002929 \times \text{PMMA} \times \text{Dip} + 0.024984 \\ & \times \text{CS} \times \text{Dip} - 2.05300\text{E}-006 \times \text{PMMA}^2 \cdots \\ & - 0.003118 \times \text{CS}^2 - 3.79682 \times \text{Dip}^2 \end{aligned} \tag{2.4}$$

2.3.4 Interaction of concentration of CS and PMMA on SWAF

The interaction of the concentration of PMMA and the CS is presented in Figure 2.10. As can be seen from the figure, the maximum SWAF was noted at a higher concentration of PMMA and CS. It could also be observed from the figure that at higher concentrations of PMMA, even though the CS concentration is low, the SWAF was found to be high. Moreover, it is interesting to note that at a lower concentration of PMMA and at a higher

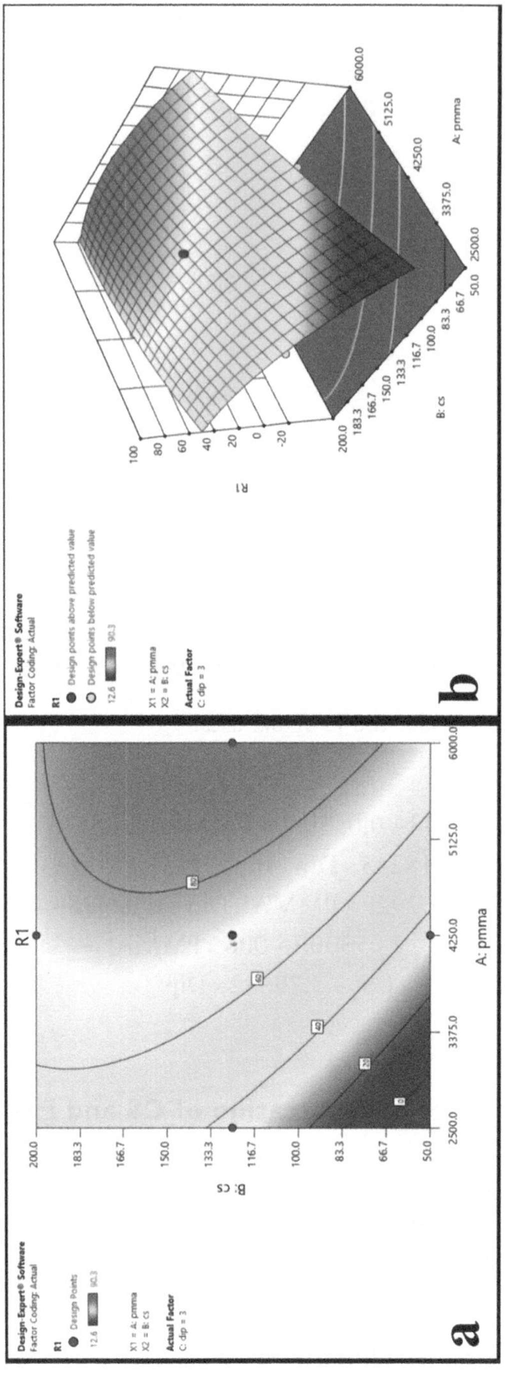

Figure 2.10 The interaction of concentration of PMMA and CS on SWAF (a) 2D, (b) Contour plot, and (c) Response surface.

concentration of CS, the SWAF noted was less. This reveals the significance of PMMA on the SWAF of the coating. The increase in the concentration of PMMA enhances the chance of binding the CS on the surface of the glass substrate. Hence at lower concentrations of PMMA, even though the amount of carbon is high, the performance noted was not very remarkable.

In addition to the concentration of PMMA and CS, it is noted from Figure 2.11 that SWAF enhances with the number of dips as well. As the number of dips increases, the layer-by-layer adhesion of CS is observed, resulting in an increase in the amount of CS, leading to an enhancement in the SWAF of the substrate. A similar trend is observed in the case of the CS and the number of dips, as presented in Figure 2.12. The SWAF of the film was found to increase with the concentration of CS and the number of dips. As it is well known that the presence of a CS and the increase in the number of dips enhances the presence of a CS in the film, the SWAF is enhanced.

As shown in Table 2.3 the PMMA concentration (A), CS concentration (B), number of dips (C), the interaction between the factors AB, BC, and AC and the second-order factors A^2, B^2, and C^2 are the effective factors that determine the response output. The p-value from Table 2.3 indicates that PMMA concentration (A) and the second-order PMMA concentration have the maximum effect on the SWAF value obtained. The observations from the response surface plots show that the effect of PMMA concentration has the highest significance compared to the other two process variables. The comparison of the response outputs obtained from optical analysis and the 3D response surface from ANOVA determines that PMMA concentration has the maximum significance.

2.3.5 Photothermal experimentation

Samples 1, 6, 9, 11, 12, and 13 were identified as the critical samples from the response of the design matrix (Table 2.2). In the selected critical samples, samples 15 and 13 exhibit the maximum solar absorption, and the minimum was observed for samples 1 and 6. Intermediate SWAF were noted for samples 9 and 11. Hence, a total of six coated samples and one uncoated sample with DI water as the working fluid were used in the experiments. Figure 2.8(a–c) shows the optical characteristics of the critical samples with (a) absorbance (α), (b) reflectance (r), and (c) transmittance (τ). The SWAF noted for samples 15, 13, 11, 9, 1, and 6 were 94%, 93%, 69%, 62%, 40%, and 32%, respectively.

Figure 2.13 gives the temperature profiles obtained for the critical CS/PMMA/epoxy coated samples along with uncoated samples. As can be seen from the profile, the temperature increases with time till noon and then decreases, following the pattern of solar radiation during daytime. For all the samples, the temperature reaches the maximum at a radiation of 950W/m^2, after which the radiation drops and subsequently the temperature of the samples. The test run T7 (corresponding to sample 7) gives a

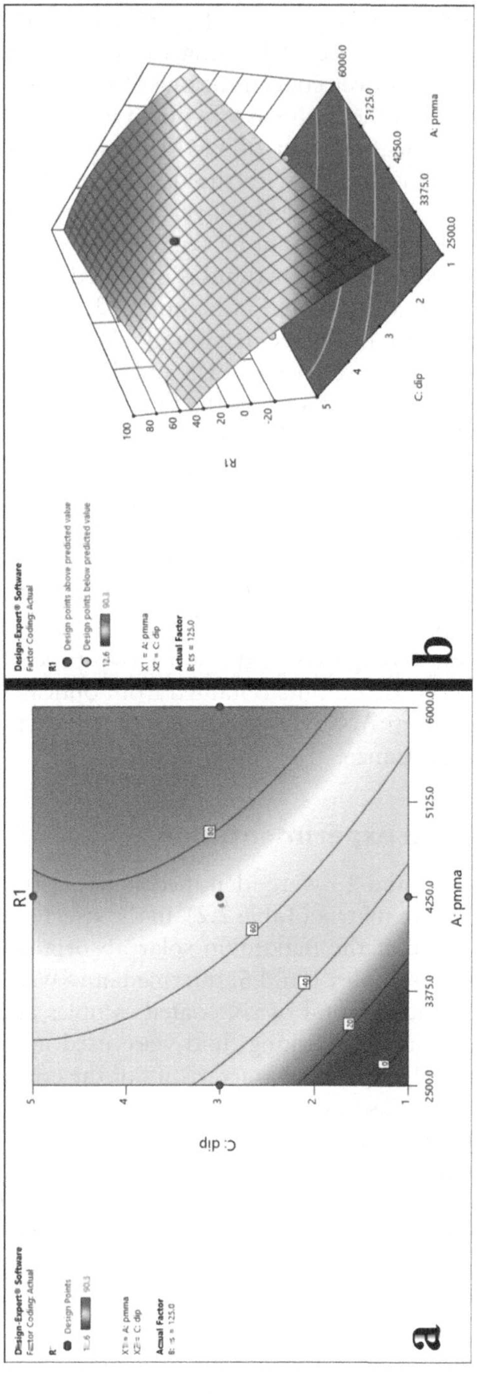

Figure 2.11 The interaction of concentration of PMMA and the number of dips on SWAF (a) Contour plot and (b) Response surface.

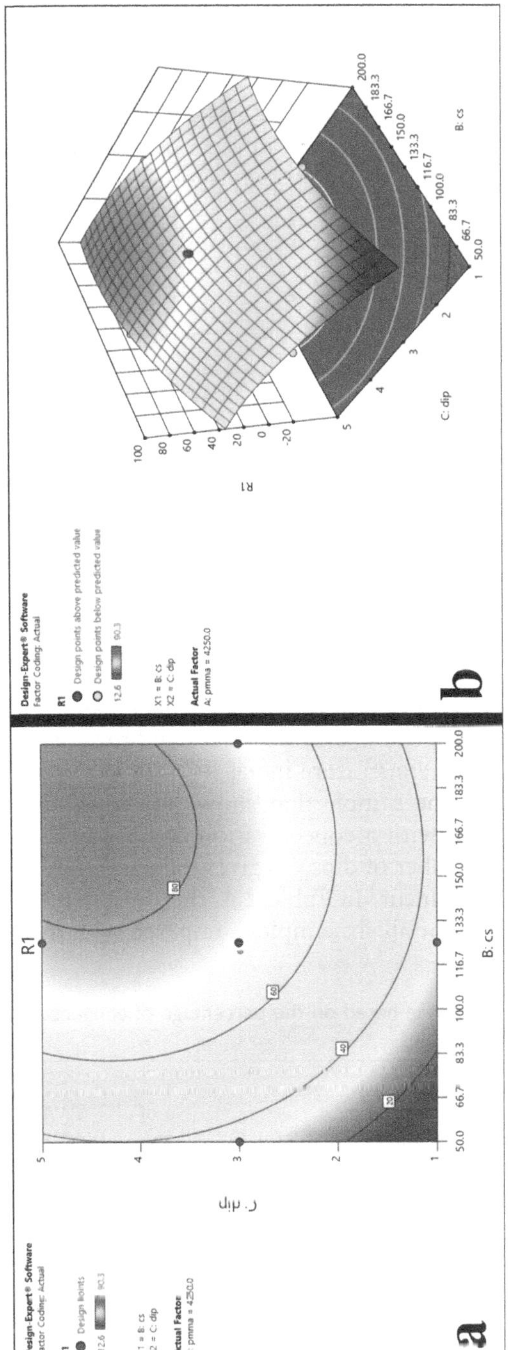

Figure 2.12 The interaction of concentration of CS and the number of dips on SWAF (a) Contour plot and (b) Response surface.

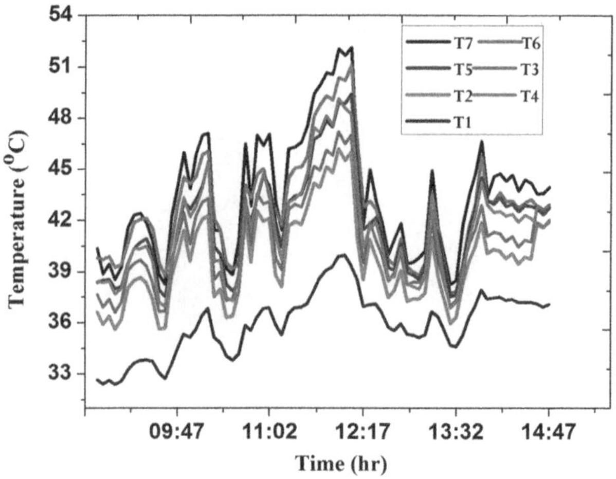

Figure 2.13 Temperature profile measured for the six selected samples using T-type thermocouples.

maximum temperature of 52.04°C, and the value noted for sample 6 (T6) was 51.31°C which is slightly less, as compared to T7. The percentage enhancement in temperature when compared to the uncoated sample is 32.35%. The minimum fluid temperatures of 46.35°C and 47.61°C were observed among coated samples for the samples T2 and T3, respectively. A careful observation of the temperature values and the corresponding SWAF could surmise that the temperature increases with the solar absorptivity of the thin film. Hence, the sample that shows the highest absorptivity, CS/PMMA/epoxy coating with a concentration of CS and PMMA as 125 mg and 6,000 mg, and number of dips = 3 gives the maximum temperature and corresponding enhancement. In Table 2.4, the percentage enhancement in temperature obtained for all six samples compared with the blank sample is

Table 2.4 Photothermal results based on the percentage of enhancement in temperature for the seven samples

Run	*Sample no*	*PMMA (mg)*	*CS (mg)*	*No. of dip*	*SWAF (%)*	*Enhancement in temperature (%)*
Run7	15	6000	125	3	94	32.35
Run6	13	5209.6	169.6	4	93	29.77
Run5	11	3209.4	169.6	4	69	25.71
Run4	9	5209.6	80.4	2	62	22.53
Run3	1	2500	125	3	40	19.42
Run2	6	3209.4	80.4	2	34	16.89
Run1	Reference run: Blank glass tube with water					

listed. From Table 2.4 it could be summarized that the maximum and minimum enhancements in temperature were obtained for samples 15 and 6, as 32.35% and 16.89%, respectively.

Run 7 and Run 6 absorb the maximum solar radiation among all the samples. As mentioned, all the hybrid absorption samples show enhanced performance when compared to direction absorption using water. As the results depict, the introduction of internal coating has improved the thermal conversion of the system. However, the efficiency could be slightly lower than the direct absorption system. The reason behind this trend is attributed to the reduced thermal losses associated with the directional absorption system. Nevertheless, for the direction absorption to be repeatable over time, the nanofluids used in the system need to be highly stable. In addition, the proposed hybrid absorption system can work on heat transfer fluid with lower solar absorptivity as well. Hence a nanofluid of lower concentration and high stability will work effectively in a hybrid absorption system.

2.4 CONCLUSIONS

The hybrid solar thermal efficiency of CS/PMMA/epoxy coating as an absorber coating was studied in the present paper. The CS used for the coating was synthesized using an APCVD setup at 950°C with C_2H_2 as the carbon precursor, and H_2 and Ar as the carrier gases. The CS coating was found to have good absorption in the UV-vis-NIR range, with the highest absorption in the UV range. The CS/PMMA/epoxy coating was prepared by dip coating. Statistical analysis was done using the DOE methodology, which helped validate the interaction effects of the three process variables, namely the concentration of CS, the concentration of PMMA, and the number of dips. The optimal process variables were developed using the CCD in RSM. The ANOVA results showed good agreement between the predicted values and the experimental data. Maximum optical absorption and temperature enhancement of 94% and 32.35% were obtained for a concentration of CS: 125 mg, a concentration of PMMA: 6000 mg, and the number of dips: 3. The ANOVA confirms that the concentration of PMMA has the highest significance among the process variables.

REFERENCES

1. Mahian, O., Kianifar, A., Kalogirou, S.A., Pop, I., Wongwises, S.: A review of the applications of nanofluids in solar energy. *Heat Mass Transf.* 57, 582 (2013).
2. Kakaza, M., Folly, K.A.: Effect of solar water heating system in reducing household energy consumption. *IFAC-PapersOnLine* 48, 468–472 (2015). https://doi.org/10.1016/j.ifacol.2015.12.423

3. Kaseman, T., Boubour, J., Schuler, D.A.: Validation of the efficacy of a solar-thermal powered autoclave system for off-grid medical instrument wet sterilization. *Am. J. Trop. Med. Hyg.* 87, 602 (2012).
4. Ghasemi, H., Ni, G., Marconnet, A.M., Loomis, J., Yerci, S., Miljkovic, N., Chen, G.: Solar steam generation by heat localization. *Nat. Commun.* 5, 1 (2014).
5. Al-Rjoub, A., Rebouta, L., Costa, P., Vieira, L.G.: Multi-layer solar selective absorber coatings based on $W/WSiAlN_x/WSiAlO_yN_x/SiAlO_x$ for high temperature applications. *Sol. Energy Mater. Sol. Cells.* 186, 300 (2018).
6. Antonaia, A., Castaldo, A., Addonizio, M.L., Esposito, S.: Stability of $W\text{-}Al_2O_3$ cermet based solar coating for receiver tube operating at high temperature. *Sol. Energy Mater. Sol. Cells.* 94, 1604 (2010).
7. Karoro, A., Nuru, Z.Y., Kotsedi, L., Bouziane, K., Mothudi, B.M., Maaza, M.: Laser nanostructured Co nanocylinders-Al_2O_3 cermets for enhanced & flexible solar selective absorbers applications. *Appl. Surf. Sci.* 347, 679 (2015).
8. Wu, L., Gao, J., Liu, Z., Liang, L., Xia, F., Cao, H.: Thermal aging characteristics of CrN_xO_y solar selective absorber coating for flat plate solar thermal collector applications. *Sol. Energy Mater. Sol. Cells.* 114, 186 (2013). https://doi.org/10.1016/j.solmat.2013.03.005
9. Li, Z., Zhao, J., Ren, L.: Aqueous solution-chemical derived $Ni–Al_2O_3$ solar selective absorbing coatings. *Sol. Energy Mater. Sol. Cells.* 105, 90 (2012).
10. Abdelfatah, A., Mohamed, L.Z., Elmahallawi, I., Abd El-Fattah, H.: Comparison of structure and solar-selective absorbance properties of Al_2O_3 thin films with Al and Ni reflector interlayers. *Chem. Pap.* 77, 5047–5057 (2023). https://doi.org/10.1007/s11696-023-02842-z
11. Selvakumar, N., Barshilia, H.C., Rajam, K.S., Biswas, A.: Structure, optical properties and thermal stability of pulsed sputter deposited high temperature $HfO_x/Mo/HfO_2$ solar selective absorbers. *Sol. Energy Mater. Sol. Cells.* 94, 1412–1420 (2010). https://doi.org/10.1016/j.solmat.2010.04.073
12. Rebouta, L., Capela, P., Andritschky, M., Matilainen, A., Santilli, P., Pischow, K., Alves, E.: Characterization of $TiAlSiN/TiAlSiON/SiO_2$ optical stack designed by modelling calculations for solar selective applications. *Sol. Energy Mater. Sol. Cells.* 105, 202 (2012).
13. Teixeira, V., Sousa, E., Costa, M.F., Nunes, C., Rosa, L., Carvalho, M.J., Collares-Pereira, M., Roman, E., Gago, J.: Spectrally selective composite coatings of $Cr\text{-}Cr_2O_3$ and $Mo\text{-}Al_2O_3$ for solar energy applications. *Thin Solid Films.* 392, 320–326 (2001). https://doi.org/10.1016/S0040-6090(01)01051-3
14. Singh, A., Kumar, M., Khullar, V.: Comprehensive modeling, simulation and analysis of nanoparticles laden volumetric absorption based concentrating solar thermal systems in laminar flow regime. *Sol. Energy.* 211, 31 (2020).
15. Panickar, R., Sobhan, C.B., Chakravorti, S.: Investigations on tribological properties of non-catalytic CVD synthesized carbon spheres in lubricant. *Diam. Relat. Mater.* 106, 107834 (2020). https://doi.org/10.1016/j.diamond.2020.107834
16. Panickar, R., Sobhan, C.B., Chakravorti, S.: Improved tribological behavior of lubricating oil dispersed with hybrid nanoparticles of functionalized carbon spheres and graphene nano. *Appl. Surf. Sci.* 540, 148402 (2020). https://doi.org/10.1016/j.apsusc.2020.148402

17. Panickar, R., Sobhan, C.B., Chakravorti, S.: Highly efficient amorphous carbon sphere-based superhydrophobic and superoleophilic sponges for oil/water separation. *Langmuir*. 37, 12501–12511 (2021).
18. Panickar, R., Sobhan, C.B., Chakravorti, S.: Chemical vapor deposition synthesis of carbon spheres: Effects of temperature and hydrogen. *Vacuum*. 172, 109108 (2020). https://doi.org/10.1016/j.vacuum.2019.109108
19. Ghanbarian, M., Nassaj, E.T., Kariminejad, A.: Synthesis of nanostructural turbostratic and hexagonal boron nitride coatings on carbon fiber cloths by dip-coating. *Surf. Coatings Technol.* 288, 185 (2016).
20. Prabeesh, P., Saritha, P., Selvam, I.P., Potty, S.N.: Fabrication of CZTS thin films by dip coating technique for solar cell applications. *Mater. Res. Bull.* 86, 295 (2017).
21. Wu, L., Yang, D., Fei, L., Huang, Y., Wu, F., Sun, Y., Shi, J., Xiang, Y.: Dip-coating process engineering and performance optimization for three-state electrochromic devices. *Nanoscale Res. Lett.* 12, (2017). https://doi.org/10.1186/s11671-017-2163-0

Chapter 3

Toward a sustainable industry from waste

Opportunities for utilization of alternative agro-residue-based raw materials

Mayand Malik, Astha Vijaivargia, Vishakha Sharma, Shailesh Pathak, and Prateek Saxena

3.1 INTRODUCTION

Agricultural residues derived from various crops, including rice, sugarcane, corn, wheat, soybean, sunflower, mustard, kenaf, bamboo, hemp, and jute, offer promising opportunities for reducing waste and utilizing resources sustainably. The wastes generated during the harvesting and processing of crops have garnered attention due to their potential applications in the production of composite materials. The promotion of resource efficiency encourages the transformation of residues into profitable goods, thereby facilitating the production of sustainable materials [1].

Agro-residues have become a substitute for traditional materials and important components of composite materials, promoting sustainability and resource optimization. The unique characteristics of various agro-residues and their compatibility with different matrixes are beneficial in producing sustainable, high-performance composites for a variety of applications [2]. The life cycle of bio-composite comprises a number of stages, starting with the extraction of energy of the sun and moving through various stages of cultivation, extraction, and manufacture. Figure 3.1 illustrates the processes associated with the conversion of renewable resources into end products.

The industrialization potential of agricultural residue has been explored in several research studies. For instance, Madurwar et al. discussed its utilization in construction materials [3], Mahawar et al. discussed its application in energy generation [4], while Rousu et al. examined its use in pulp production for papermaking [5]. Lee et al. investigated its potential as a source of biofuel [6], and Pratiwi et al. explored its suitability for packaging purposes [7]. Additionally, Sanyang et al. discussed the utilization of agricultural waste in the creation of composites [8].

This chapter aims to explore the potential of various agricultural residues for the production of sustainable composites. The chapter provides an in-depth understanding of commercially viable new product development that utilizes agricultural waste, discussing the decision-making criteria for their raw materials based on their morphological, chemical, and mechanical

DOI: 10.1201/9781003527275-3

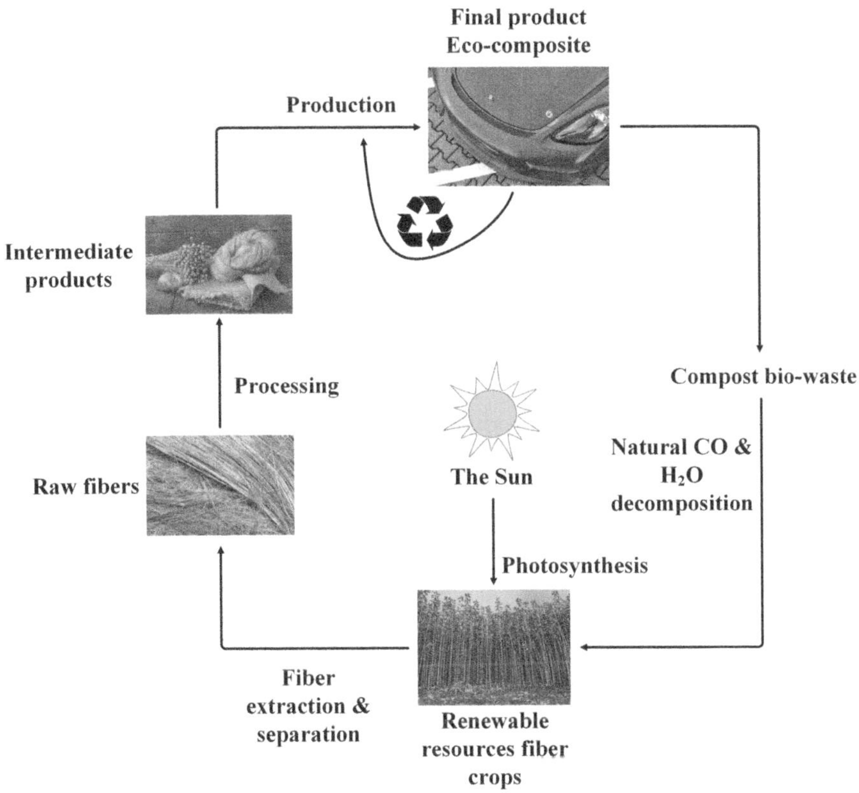

Figure 3.1 Life cycle of bio-composites.

properties. The lack of complete understanding of current activities and their environmental consequences is an urgent concern. Therefore, the purpose is to conduct a detailed analysis, with a particular focus on the use of agricultural residue as a viable strategy to promote environmental sustainability.

3.2 CONVERSION OF RAW MATERIALS

The recent attention surrounding the conversion of crop residues into finished products is attributed to their potential for sustainable resource utilization and the development of products with enhanced value. Various crop residues, including those from rice, sugarcane, corn, wheat, soybean, sunflower, mustard, kenaf, bamboo, hemp, and jute, necessitate transformation into finished products through the conversion into various materials suitable for a variety of applications. This procedure addresses the urgent demand for sustainable practices in industry and agriculture as well as contributing to a circular economy.

Crop residues are transformed into final products using a variety of methods and technologies. To remove or alter their components, these are subjected to processes like drying, grinding, chemical, and thermal processing. The finished products are subsequently adopted for a variety of applications, such as construction, packaging, textiles, biofuels, and composites. Various industries, including automotive, aerospace, and packaging, have the potential to benefit from the use of composites fabricated using crop residues [9].

3.2.1 Extraction and conversion techniques of rice straw

Rice straw, a significant by-product in the agricultural sector, exhibits considerable potential for conversion into valuable products through multiple techniques. Thermochemical conversion and anaerobic digestion are two common processes used to convert rice straw.

3.2.1.1 Thermochemical conversion of rice straw

By applying thermal processes, biomass is transformed into various forms of energy, such as heat, electricity, or biofuels. This requires the degradation of biomass through heat exposure, often above 300°C. Direct combustion, gasification, and pyrolysis are all types of thermal conversion processes. Rice straw is a useful feedstock for thermal conversion because of its substantial energy content and relatively high carbon and hydrogen concentrations. The distribution and quality of the manufactured products are affected by several process variables, including temperature, reaction time, heating rate, oxygen presence or absence, catalyst use, and pressure.

3.2.1.2 Applications of rice straw

The cultivation of multiple crops in the same area annually gives rise to challenges associated with soil incorporation. This occurs as a result of the straw having insufficient time to disintegrate, which leaves it with poor soil fertilization properties and hinders the development of crops. As a result, despite being outlawed in the majority of rice-growing nations due to emissions and the associated health risks, open-field straw burning has increased dramatically over the past ten years. As a result, it is critical to look for sustainable ideas and solutions that can lessen their environmental impact while enhancing the value of the rice production system. Figure 3.2 shows various management options for rice straw.

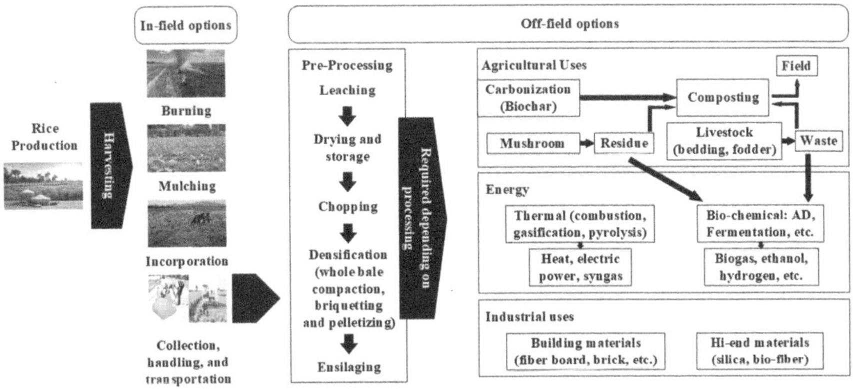

Figure 3.2 Rice straw management options.

3.2.1.3 Applications of rice residues in 3D printing

Traditional plastic storage boxes can be replaced with biodegradable storage materials made using rice residues. These materials are 3D-printed in various shapes and sizes. Eco-friendly construction materials like bricks and blocks with high sustainability are created using rice residues. Being biocompatible and biodegradable, rice residues can be used to 3D print medical implants, such as bone implants. A variety of consumer products, such as phone cases, sunglasses, and toys, are printed with custom designs and shapes by utilizing rice residues [10]. The matrix material required in 3D printing rice residue composites is a thermoplastic or thermosetting polymer. To create a homogeneous blend of rice residue and matrix material, the residue is first cleaned, dried, and then ground into a fine powder. Further residue powder is mixed with the matrix material, and the resulting composite material is printed using fused filament fabrication (FFF). The content of powder mixed with the matrix depends on the desired properties [11].

3.2.1.4 Rice straw and rice husk composites

Ceramic composites such as rice husk ash-filled cement and rice husk-filled gypsum wall and ceiling panels have employed rice husk as a reinforcing material. The rice straw composite business uses high-density fiberboards, medium-density fiberboards, and particleboards. The use of rice straw in the construction of furniture, car doors, and roofing is widespread. Composite materials made of polypropylene and polyvinyl chloride incorporate rice straws as reinforcing fibers. Over recent years, there has been notable progress in the advancement of rice residue fiber-reinforced polylactic acid

(PLA) composites. The enhanced performance of composites is attributed to these applications. The study conducted by Yussuf et al. [12] found that the interface strength of the rice residue (RR)/PLA composite exhibited a significant weakness, characterized by the presence of numerous voids and occurrences of fiber pullouts. However, the application of a polymer coating on the surface of the fiber resulted in enhanced bonding properties, thereby preventing any potential damage to the fiber during the processing stage. In order to enhance the interface performance of rice straw (RS) and PLA, several chemical and physical treatments can be employed. These treatments include maleic anhydride polypropylene (MAPP) and MAPE modification, as well as alkali treatment, silane coupling agent treatment, crosslinking agent treatment, glycidyl methacrylate (GMA) modification, acrylic acid (AA) modification, isocyanate (MDI), and glycidyl methacrylate (GMA) modification. Table 3.1 presents a comprehensive compilation of the mechanical properties, including tensile strength (TS), tensile modulus (TM), flexural strength (FS), and flexural modulus (FM), exhibited by RR/PLA composite materials that have been subjected to different treatment methods. The TS of RH/PLA composites, which underwent treatment with MAPP and polyethylene glycol (PEG), exhibited the lowest value, measuring approximately 25 MPa. The highest TS was observed in RH/PLA composites that underwent treatment with silane, MMA, and ATP, measuring approximately 60 MPa.

In order to fabricate nanocomposites with exceptional performance in food packaging, biomedical, drug delivery, wound dressing, and biosensor applications, RS nanocellulose in powder form is added to epoxy resin and mixed well. Natural rubber composites reinforced with RS fiber are used to make mats, toys, washers, and the interior of cars and trains [21].

Table 3.1 Mechanical properties of RR/PLA composites

TS (MPa)	*TM (GPa)*	*FS (MPa)*	*FM (GPa)*	*Mark*	*Year*	*References*
—	—	85	4	—	2010	[12]
56	4.6	83	5.8	30% straw	2011	[13]
51.8	—	—	—	GMA	2013	[14]
36.7	4.1	75.4	4.47	20%	2015	[15]
27	—	—	—	RS 25%, PEG	2017	[16]
28.39	—	—	—	RS 10%	2019	[17]
50	—	—	—	MDI 20% rice husk (RH)	2020	[18]
38	—	75	—	30% RS	2021	[19]
59	0.55	92	3.25	A + U	2021	[20]

3.2.2 Extraction and conversion techniques of bagasse

Through the implementation of diverse methodologies, it is possible to effectively convert sugarcane bagasse, a significant residual product derived from the sugar industry, into valuable products. Thermochemical conversion, biochemical conversion, and physical treatments are a few conversion processes for sugarcane bagasse.

3.2.2.1 Thermochemical conversion of bagasse

Sugarcane bagasse is converted using thermochemical methods such as pyrolysis, gasification, and combustion. Pyrolysis is the process of heating biomass in the absence of oxygen to create syngas, charcoal, and bio-oil. Syngas generated during pyrolysis or gasification is further processed into value-added products and used to generate heat and power.

3.2.2.2 Applications of sugarcane bagasse

Bagasse is used in the production of fuels, fiber products (such as paper and cardboard), concrete construction materials, animal feed products, and chemicals like furfural. Energy sources, such as electricity generated by combustion or gasification, are also processed using bagasse.

3.2.2.3 Applications in composites

A green composite made of tapioca starch and sugarcane fiber cellulose (SCFC) is used to make disposable food packaging. Glycerol is utilized as a plasticizer to enhance flexural properties. The composite is safe for microwaves, freezers, and multipurpose containers. This measure is anticipated to enhance the manufacturing output of environmentally friendly products while concurrently reducing the demand for paper and paper-related materials. Details about various polymer matrix and bagasse fiber combinations are provided in Table 3.2.

3.2.2.4 Applications of bagasse residues in 3D printing

By cutting sugarcane bagasse fibers into the proper sizes and combining them with polymer matrix materials like polypropylene or recycled polypropylene (rPP), sugarcane bagasse fibers are used as reinforcement material in composites fabricated by 3D printing. Mixing fibers with polymer matrix material improves the mechanical properties, such as strength and stiffness, of the overall printed part. Sugarcane bagasse is also used to produce bio-based polymer matrix material in filament form, which is further used in 3D printing composites. These filaments are produced by extruding sugarcane

Table 3.2 Mechanical properties of bagasse residues

S. No.	*Composites formed*	*Mechanical properties*	*References*
1	Polyvinyl alcohol with nanocellulose from bagasse	After 5% nanocellulose weight fraction, TS dropped	[22]
2	Bagasse (20% weight fraction) with polypropylene	Compared to neat resin specimens, TS, modulus, FS, and impact strength increased by 15%, 50%, 35%, 32%, and 45%, respectively	[23]
3	Polyester-sugarcane bagasse continuous, aligned fibers with volume fraction 10, 20, and 30%	Fiber volume fraction linearly increased impact energy	[24]
4	Bagasse-phenolic resins. Samples were made using the hand lay-up technique	Sugarcane bagasse fibers minimize composite water content and void	[25]
5	Compression-molded polyester composites using 5% sugarcane bagasse	Esterified bagasse fibers bonded with polyester had 71% higher TM than neat polymers	[26]
6	Vacuum-bagged bagasse (10, 15, 20% volume fraction) untreated and alkali, acrylic acid-treated polyester composites	Acrylic acid-treated fiber-reinforced composites with water resistance have higher tensile and flexural strengths	[27]
7	Twin screw extrusion and injection molding reinforced polypropylene-reinforced alkali-treated bagasse fibers. Fiber weight fraction ranged from 5% to 30%	25% fiber-reinforced composites have a 1460 MPa TM versus 900 MPa for polymer only. similar loaded composites had an 1800 MPa FM, far higher than the control specimen	[28]

bagasse fibers of a suitable size and shape so that they can be used as feedstock in a 3D printer.

3.2.3 Extraction and conversion techniques of corn residues

The components of corn residue can be extracted and turned into useful products using a variety of conversion procedures. These methods typically include treating corn residue physically, chemically, or biologically to unravel its intricate structure and liberate its constituent parts, including cellulose, hemicellulose, and lignin. Figure 3.3 shows the various ways that

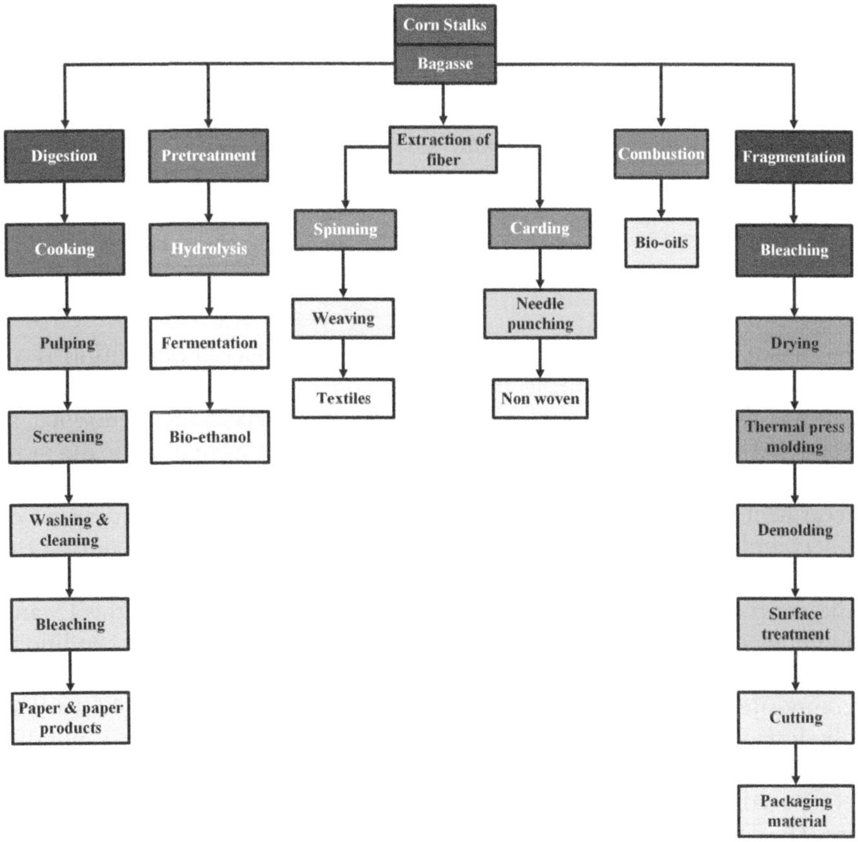

Figure 3.3 Corn stalks and bagasse applications.

corn stalks and bagasse, two common agricultural by-products, can be used in sustainable ways.

3.2.3.1 Thermochemical conversion of corn residues

Using thermochemical conversion methods is a potential way to turn corn residue, often known as corn stover, into useful products. These processes involve heating the biomass and triggering chemical reactions to produce biofuels, biochemicals, or other high-value products.

3.2.3.2 Composite materials

The limited compatibility between natural fibers and matrix materials is a widely acknowledged drawback, primarily attributed to the hydrophilic

Table 3.3 Impact of fiber treatment on the tensile strength and tensile modulus of CF/PLA composites

Composite	*Average tensile strength (MPa)*	*Average tensile modulus (GPa)*
UCF/PLA	40.29	39.74
ACF/PLA	47.52	43.36
SCF8/PLA	53.4	51.68
SCF12/PLA	54.31	53.1

nature of natural fibers. Hence, the process of surface modification is an essential requirement during the preparation of composites consisting of natural fibers and polymer materials. Among the methods mentioned in Section 3.2.1.4, alkali treatment and silane treatment have become the most common and effective techniques. The application of sizing agents in surface treatment offers several advantages. It improves the handling properties of fibers and provides protection against fuzzing and fragmenting during the processing of composites. Additionally, it improves the wetting of fibers by the matrix material. The data presented in Table 3.3 illustrates the TS and modulus of corn fiber (CF)/PLA composites at varied interface conditions and varying sizing concentrations. The untreated corn fiber (UCF)/PLA composite shows a TS of 39.8 ± 1.4 MPa and a TM of 4.97 ± 0.34 GPa. The alkali-treated carbon fiber (ACF)/PLA composite demonstrates a notable enhancement of 18.1% in TS, along with a 9.5% improvement in TM, when compared to the untreated unidirectional carbon fiber (UCF)/PLA composite. The TS of sizing-treated (SCF)12/PLA (12% NaOH by weight) is measured to be 49.5 ± 2.0 MPa, while its TM is found to be 6.63 ± 0.07 GPa. These values are 34.7% and 33.4% higher, respectively, compared to the corresponding values of UCF/PLA [29].

3.2.3.3 *Applications of corn residues in 3D printing*

Corn residues, similar to rice residues, are also utilized to 3D print sturdy, long-lasting parts for the automobile and aerospace industries. The corn residues are first collected, cleaned, and then either ground to a fine powder or chopped into small pieces suitable for 3D printing. Further steps are the same as those in rice residues to fabricate composite material via 3D printing.

3.2.4 Extraction and conversion of wheat residues

The productive use of wheat residue presents a multitude of potential applications, including energy generation, biofuel production, and the manufacturing of composite products, thereby preventing its disposal as waste. Various extraction and conversion processes have been developed to effectively utilize wheat residue and mitigate its adverse environmental impacts.

3.2.4.1 Thermochemical conversions

Thermochemical techniques are employed to convert wheat residue, wherein heat is applied to induce chemical reactions that yield valuable products such as biofuels, charcoal, and syngas. Commonly used methods include pyrolysis, gasification, and torrefaction. The potential of these methods for the conversion of wheat residue has been thoroughly investigated [30]. As shown in Figure 3.4, the processes used for making final value-added products from both wheat and RS are quite comparable.

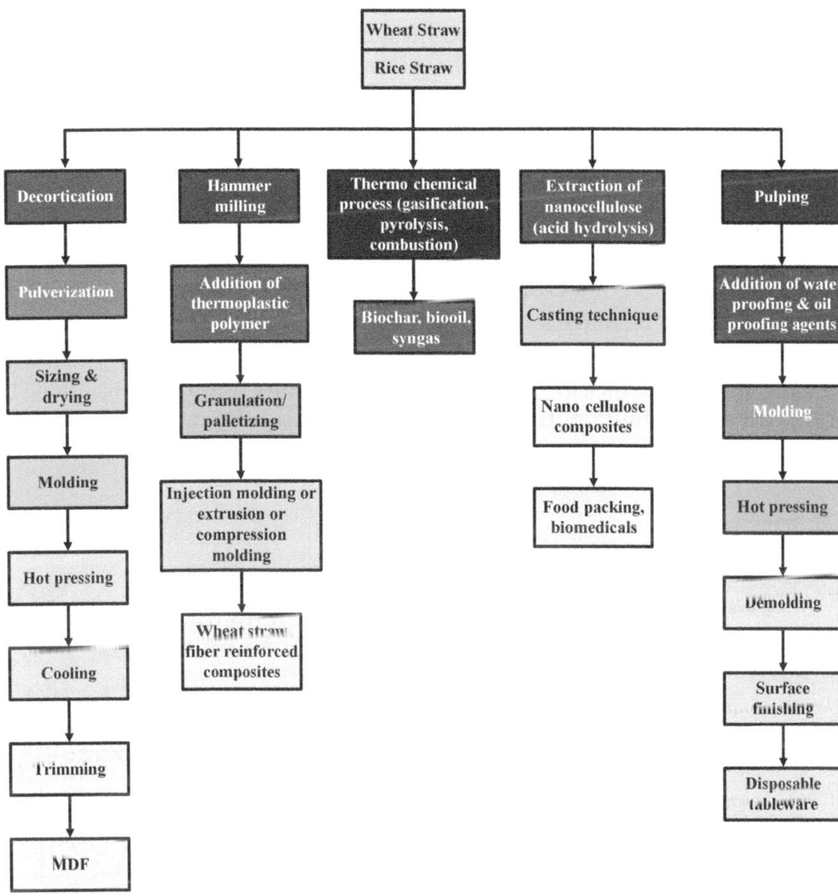

Figure 3.4 Wheat and rice straw applications.

3.2.4.2 *Applications of wheat residues*

Wheat straw is a great source of cellulosic fiber and is used as filler and reinforcement in both structural and nonstructural composites. The automobile, building, housing, and furniture industries all benefit from the use of composites made from wheat straw [31]. Fiber derived from the cellulose and hemicellulose of wheat straw serves as a valuable resource for various applications, including pulp and paper production, adhesives and binders, biopolymers such as polysaccharides, and composites. Wheat straw composites possess the ability to be utilized in automotive interiors for both structural and semi-structural applications. The quarter-trim bin utilized in Ford Motor Company vehicles is constructed from a composite material derived from wheat straw. The utilization of wheat straw fibers was explored in a study investigating their compatibility with biodegradable and aromatic co-polyester materials. The objective was to develop a fully sustainable composite material suitable for non-food packaging and other agricultural applications with a limited lifespan. To achieve this, hemicellulose sugars were extracted from the wheat straw fibers through acid hydrolysis after fractionation.

3.2.4.3 *Manufacturing of composites*

Wheat straw is harvested, cleaned, and reduced in size before being chemically treated. A matrix material is then chosen, and the treated straw and treated matrix are combined. The mixture is then molded and allowed to cure. In order to produce a composite with better qualities, the reinforcing wheat straw fibers are combined with a compatible matrix substance, such as a thermoset polymer or thermoplastic [32]. Figure 3.5 illustrates the general process of manufacturing composite materials using wheat straw.

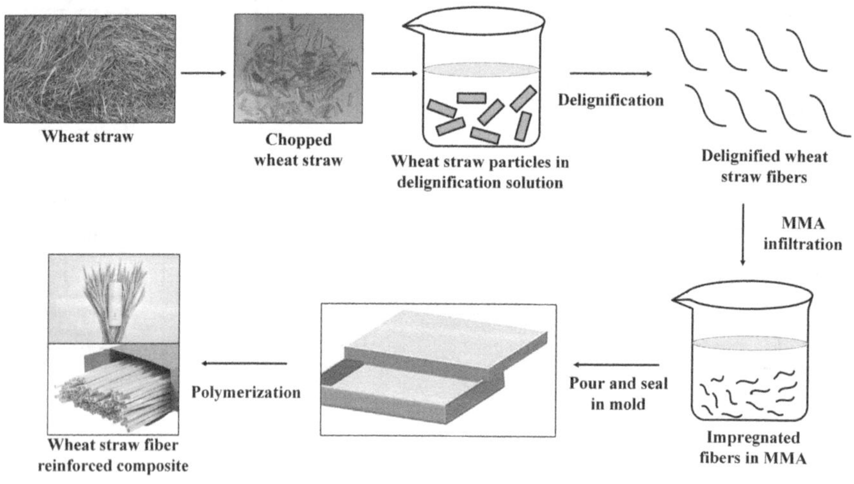

Figure 3.5 Production of composite materials reinforced with wheat straw fibers.

Table 3.4 Mechanical properties of WSF reinforced composites

Composite	*Tensile strength (MPa)*	*Tensile modulus (GPa)*	*References*
PP/WSF (mechanically processed)	40.8	2.99	[34]
PP/WSF (chemically processed)	35.5	2.44	[34]
PP/WSF	21.8	1.673	[35]
HDPE/WSF	20.8	3.1	[36]
PLA/WSF	56.52	4.5	[33]

Table 3.4 provides a brief review of the TS and TM values exhibited by composites made of polymer and wheat straw fiber. Panthapulakkal and Sain utilized polypropylene and high-density polyethylene (HDPE) as the polymer matrix for the manufacture of a composite material incorporating wheat straw fibers (WSF) [32]. Nyambo et al. employed PLA as the matrix material [33].

3.2.4.4 *Applications of wheat residues in 3D printing*

Utilizing residues from the processing of wheat, various small parts for agricultural machinery, like those used in tractors, can be 3D-printed. Composite filaments are made using wheat residues by mixing the residue with polymers like PLA, PET, etc. Residue as a reinforcing filler material improves the overall mechanical and thermal properties of the composite. Wheat residues are also combined with other natural fibers like flax, hemp, or bamboo to make composites with improved mechanical properties [37].

3.2.4.5 *Manufacturing of tableware and drinkware*

A wide range of raw materials are available for the production of biodegradable and compostable tableware. The detailed procedure for producing tableware using wheat straw as the main raw material is shown in Figure 3.6

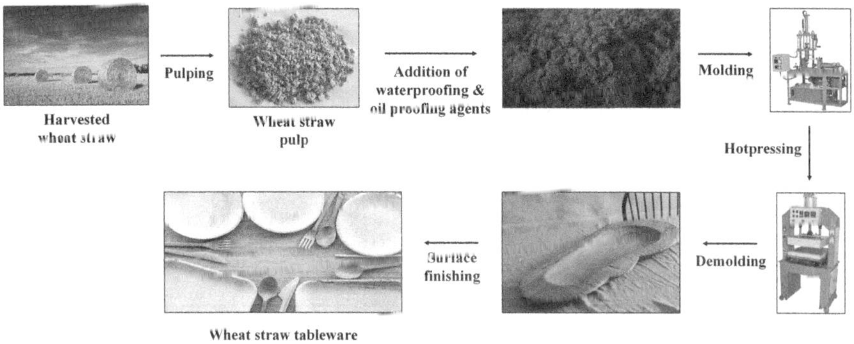

Figure 3.6 Manufacturing process of disposable tableware utilizing wheat straw.

3.2.5 Extraction and conversion techniques of soybean residues

Soybean residue, commonly known as soybean meal or soybean cake, is derived as a solid by-product subsequent to the extraction of soybean oil. This residue is further processed to extract or convert its components for use in a variety of applications. It is a valuable source of protein, fiber, and other nutrients. To make effective use of soybean residue, a variety of extraction and conversion procedures are used.

3.2.5.1 Applications of soybean residues in composites

Fuel efficiency and service life are key sustainability factors for automobiles, and the selection of materials significantly influences their performance. Extensive research has been conducted on the utilization of natural fiber soybean in composite applications. The incorporation of soy protein (SP) into PLA composites has the potential to significantly decrease overall expenses while simultaneously enhancing the rate of breakdown. Nagarajan et al. conducted a study to examine the impact of integrating 30 wt% lignocellulosic fibers into a biodegradable polymer matrix consisting of a pre-blend of poly (3-hydroxybutyrate-co-3-hydroxyvalerate) (PHBV) and polybutylene adipate-co-terephthalate (PBAT) [38]. The tensile properties of composites consisting of soybean residue and polymer are presented in Table 3.5.

3.2.6 Extraction and conversion techniques of kenaf residues

Among various extraction and conversion techniques, biochemical conversion processes, such as anaerobic digestion and fermentation, have the capability to produce biogas and biofuels. Pyrolysis and gasification are utilized to produce biochar, bio-oil, and syngas. Biorefining enables the fractionation of kenaf biomass into valuable components suitable for the production of biofuels, bioplastics, and biochemicals.

Table 3.5 Mechanical properties of soybean composites

Composite	*Tensile strength (MPa)*	*Tensile modulus (GPa)*	*References*
PHBV/PBAT/Soy stalk	22.44	1.95	[38]
PLA/SP composite	44.7	4.17	[39]
PLA + 30% soy stalks	57.04	3.35	[13]

3.2.6.1 *Applications of kenaf residues*

Kenaf is a high-yielding fiber crop used in many manufacturing applications as a raw material, ranging from conventional garments, yarns, and ropes to modern construction materials, composites, and lightweight automobile components. Figure 3.7 depicts various applications of kenaf residue and the associated processing techniques.

a) **Kenaf fiber composites**: Natural fibers like kenaf outperform glass fiber in pricing, density, renewability, recyclability, abrasiveness, and biodegradability. Lightweight and recyclable natural fiber composites are utilized in automobiles. Thermoformed polypropylene fibers are employed in some applications to exert compression on kenaf and jute fibers. Table 3.6 displays the most recent NFPC applications for aircraft engineering [40, 41].
b) **Food and non-food packaging material**: The utilization of kenaf/plastic composite pellets is observed in many applications, including commercial food storage containers, non-food packaging such as disposable packaging, as well as bulk pharmaceutical medication and chemical packaging. Moreover, Kenaf finds application in the

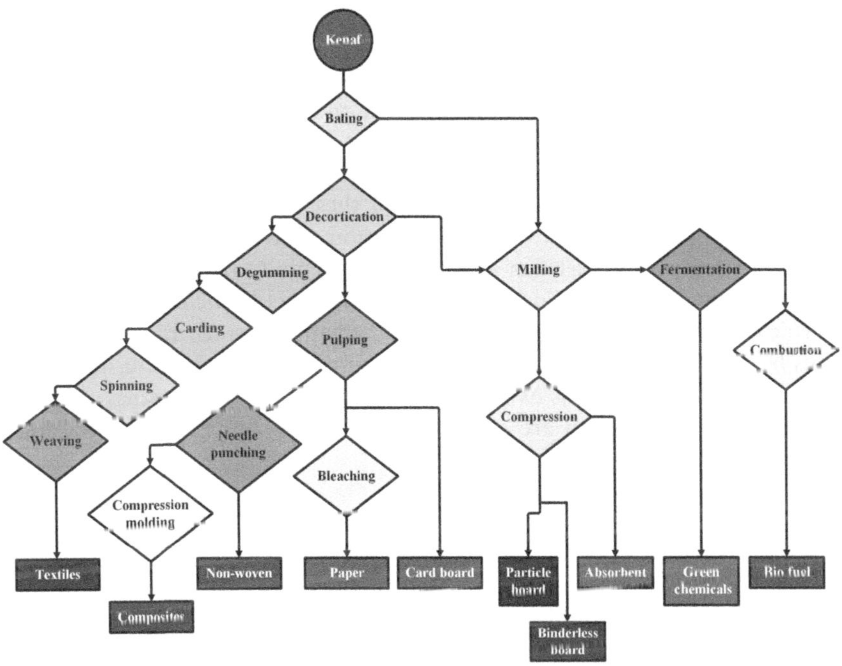

Figure 3.7 Kenaf applications.

Table 3.6 Natural fiber polymer composites' (NFPCs) applications in aircraft engineering

Types of NFPC	*Applications in aerospace*
Natural fiber-based sandwich composites	Cabin sidewalls
Natural flax fiber hybrid composite	X3 high-speed demonstrator
Cotton fiber-reinforced bio-based Polyurethane composites	Aircraft interiors
Kenaf fiber or hybrid kenaf fiber	Radome components
Natural fiber polymer composites	Seats (seat pan, armrest, and backrest), flooring, and decking
Natural fiber-reinforced thermoset sandwich panels from flax fabric-phenolic skins and Nomex honeycomb core	Ceilings, galley walls, lavatories, and cargo hold liners
Natural fiber-reinforced panels	Floors, sidewalls, bulkheads, stowage compartments, and other cabin parts

production of packaging components utilized in the assembly of large household items such as refrigerators and electronic devices.

3.2.7 Extraction and conversion techniques of bamboo residues and their applications

Various techniques such as gasification, pyrolysis, and combustion are employed to transform bamboo residues into biofuels, biogas, or heat energy. The bamboo residue is utilized in the production of bio-composite materials. The manufacturing process of these materials entails the treatment of by-products and their combination with other organic constituents. Bio-composites, serving as an environmentally friendly alternative to conventional materials, find utility in multiple fields such as construction, furniture manufacturing, and various other applications. Bioactive chemicals like lignin, flavonoids, and phenolic compounds are present in bamboo residues. These chemicals exhibit antioxidant, antibacterial, and other beneficial properties, which are useful in fabricating composites for the food, medicine, and cosmetics industry [42].

3.2.7.1 Applications of bamboo residues in composites

The study conducted by Tan et al. examined the pertinent relationship between the structure and properties of bamboo particles and bamboo thermoplastic composites (BPCs). The study involved the compounding of six different types of bamboo particles, namely bamboo culm (BC), bamboo green (BG), bamboo timber (BT), bamboo yellow (BY), bamboo node (BN),

Table 3.7 Mechanical properties of bamboo residue composites

Composite	*Tensile strength (MPa)*	*Tensile modulus (GPa)*	*References*
BC/PP	54.8	4.78	[43]
BC/LDPE	11.9	0.523	[44]
BC/20% polyester	52.15	—	[45]

and bamboo processing residual (BR), with polypropylene (PP) in order to produce bamboo-polypropylene composites (BPCs). Table 3.7 displays the mechanical properties of composites fabricated using bamboo residues.

3.2.8 Extraction and conversion techniques of hemp residues

Hemp fibers are separated and refined for use in textiles and building materials using fiber extraction procedures, including retting and decortication. Additionally, hemp residues are bio-converted using microbial or enzymatic processes into biofuels, biochemicals, or bioplastics. Although further study and development are needed to fully explore their possibilities, these methods present opportunities for using hemp residue in an efficient way. Figure 3.8 shows the sequential steps needed to extract hemp fibers and husks from unprocessed hemp straw.

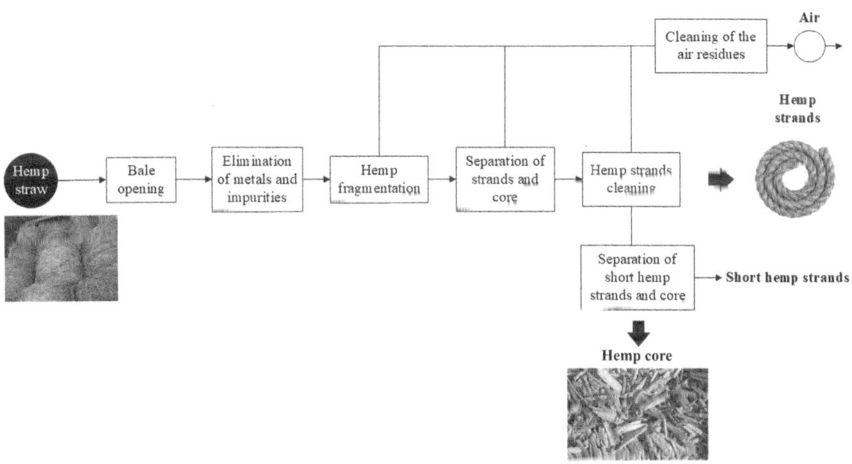

Figure 3.8 Hemp straw processing to obtain hemp strands or fibers and hemp cores or hurds.

3.2.8.1 Applications of hemp residues

In the textile, luxury retail, food, beverage, furniture, construction, automotive, cosmetics, and personal care industries, hemp offers a wide range of potential applications.

a) **Composites**: Hemp fibers are employed as a substitute for synthetic fibers to provide reinforcement in composites with polypropylene and epoxy resin as matrixes. Bast fibers are highly suitable for the fabrication of composite and bio-composite materials, owing to their considerable length and strength. Hemp seeds, whether in their modified or unmodified form, can also serve as a viable source for the production of building materials. Hemp fiber-reinforced polymer composites are used in making window and floor coverings [46].

3.2.9 Processing methods for hemp composites

a) **Hand lay-up**: The hand lay-up technique is a frequently employed method for fabricating various bio-composites. This technique involves placing the fibers in the mold and using rollers to apply resin to the fibers. After that, a vacuum process is typically employed for curing. Some of the primary advantages of this method include its simplicity, low processing costs, and ability to generate complex designs. Nevertheless, the protracted duration and substantial labor involved are among the limited disadvantages of this process.
b) **Resin transfer molding**: Resin transfer molding has a high production rate and is economical. After placing the fibers, the resin material injection is done in the closed cavity mold first. The curing process is subsequently conducted under reduced vacuum pressure, resulting in completion within a timeframe of 30 to 60 minutes, facilitated by the utilization of vacuum. After that, the fibers are added to the mold before the resin is injected. The fibers will then impregnate the resin after injection. Process variables like injection pressure, vacuum pressure, fiber content, and temperature have an impact on the composite's ultimate mechanical and physical properties.
c) **Pultrusion**: Jute fibers are impregnated with resin material during the production of jute polymers and composites, and the material is subsequently passed through a heated die. It is challenging to maintain fiber orientation throughout this procedure. However, the method is renowned for producing intricate geometries and thin cross-sectional structures with high automation. The utilization of this particular technique results in the production of jute products that consist of a significant proportion, up to 70%, of jute fibers.
d) **Extrusion**: Single or twin screws, capable of both clockwise and anticlockwise rotation, are utilized in this technique. Single-screw extruders are commonly utilized in situations, where minimal fiber

and matrix material mixing is required, primarily due to their limited mixing capabilities. Contrarily, a twin screw extruder is renowned for a superior mixing effect and high thrust forces, which help to evenly distribute fibers throughout the composite.

3.3 DRAWBACKS/MAJOR CHALLENGES

Agricultural residues possess the potential to be utilized in different ways, thereby facilitating waste reduction and fostering the promotion of sustainable resource utilization. Nevertheless, the utilization of such natural by-products is not without drawbacks and poses considerable challenges. The variability in the composition and quality of crop residue poses a substantial concern. The composition and energy content of residues may vary depending on factors such as crop type, cultivation techniques, weather conditions, and harvesting techniques [47]. The difficulties and expenses involved in gathering, storing, and transporting crop residue present another difficulty. The extraction and management of large amounts of residue impose significant demands on both human resources and financial capital, particularly in remote or rural areas [48].

The utilization of agricultural residues in the production of composites by 3D printing is advantageous due to their abundant availability, renewable characteristics, cost-effectiveness, reduced environmental footprint, lightweight composition, high structural integrity, biodegradability, tailorable properties, and potential for localized manufacturing. Lu Chen et al. presented a study on a three-dimensional (3D) printed evaporator made of a porous wood-mimetic cellulosic composite. This design was inspired by the distinctive liquid transportation properties of natural wood and aimed to achieve long-term salt rejection. The first elements for the manufacture of printable ink consisted of cellulose and carbon black. The solid particles and fibrils were dispersed and achieved a stable and homogeneous suspension through the process of sonication in water. The cellulose nanofiber (CNF)/carbon black slurry, which had a solid content of around 2 wt%, was successfully extruded from a fine nozzle in a continuous manner. This resulted in the formation of a long line measuring 8 cm, thus indicating the favorable printability of the slurry [49]. Varying physical and chemical properties exhibited by crop residues pose challenges in achieving consistent and standardized feedstock materials for 3D printing. In order to facilitate the production of uniform composite materials suitable for printing applications, additional preprocessing steps such as grinding, sieving, or blending are required. One more challenge arises from the compatibility between agricul tural residues and the 3D printing process. Crop residues often contain natural fibers, lignocellulosic materials, and other organic constituents. These components can significantly influence the rheological properties and printability of such materials. The attainment of optimal flow characteristics, printability, and interlayer adhesion poses challenges when utilizing composites created from crop residue. In order to improve the printability and mechanical properties of composite compositions, it is essential to formulate and optimize them appropriately through the addition of binders or additives [50]. To enhance the mechanical properties of composites fabricated using crop residues, it is necessary to employ reinforcement techniques such as the incorporation of synthetic or natural fillers, fibers, or resins [40].

3.4 CONCLUSION

The extraction and conversion of agricultural residues significantly contribute in enhancing resource efficiency, mitigating waste generation, and facilitating the production of environmentally sustainable products and materials. Various agricultural residues can be effectively transformed into valuable products, including biofuels, biochemicals, bioplastics, animal feed, organic fertilizers, and energy sources, using both thermochemical and biochemical conversion techniques. However, there are several challenges associated with residue heterogeneity, collection, and transportation costs, as well as concerns regarding soil fertility and ecosystem services. Therefore, simplifying residue collection, developing new technologies, and ensuring sustainable practices are essential in order to fully utilize the potential of agricultural residues and pave the way for a better and more responsible future. Furthermore, there are additional challenges associated with the use of agricultural residue in 3D printing, such as the need for preprocessing to obtain consistent feedstock materials and the suitability of organic constituents for the printing process. The optimization of crop residue-based composites is essential, and the incorporation of additives is necessary to enhance printability and mechanical properties. These initiatives demonstrate the ability to promote a more sustainable path by reducing waste and enhancing the efficiency of resources within a variety of industries such as construction, packaging, textiles, biofuels, and composites.

3.5 CASE STUDIES

3.5.1 Transforming agricultural waste into biodegradable plastics [51]

Plastics manufactured from petroleum-based sources, being one of the major causes of environmental contamination, have presented substantial difficulties for ecosystems and human well-being on a global scale. Ukhi India, an innovative company, has developed to address these concerns by developing new methods to utilize agricultural waste as a valuable resource for producing biodegradable plastics that can be reinforced with natural fibers like basalt, kenaf, hemp, jute, and bamboo to produce eco-friendly composites with a high strength-to-weight ratio.

Ukhi India provides a variety of environmentally friendly plastic granules made from agricultural waste that are biodegradable and compostable. These granules are utilized in the production of several flexible packaging options for disposable plastic items, including shopping bags, trash can liners, e-commerce packaging, and industrial wraps. Their materials, unlike traditional plastics, undergo decomposition into carbon dioxide and water when they contact the soil. This process eliminates the presence of any harmful residues and prevents their persistence in the environment for a long period of time. The natural breakdown process greatly diminishes the environmental impact of disposable plastics.

3.5.2 Turning papaya peel waste into revolutionary nanocomposites: Combating pollution and infections [52]

A team of international researchers has successfully created a fluorescent carbon dot and reduced graphene oxide (RGO) nanocomposite from papaya peel waste, marking a significant advancement in sustainable environmental management and antimicrobial methods. The research, published in the Journal of Bioresources and Bioproducts, describes a single-step hydrothermal process for creating the nanocomposite. The research was motivated by the necessity to tackle water contamination and environmental damage resulting from the widespread utilization of organic chemicals, specifically in the agricultural sector. The objective of the study was to create a photocatalyst with the ability to break down these harmful substances when exposed to sunlight. The nanocomposite exhibited remarkable antibacterial efficacy against Gram-positive Bacillus subtilis and Gram-negative Pseudomonas aeruginosa microorganisms. The results of the study emphasize the creative utilization of agricultural waste in the creation of products with diverse applications. The nanocomposite's ability to operate as both a photocatalyst and an antimicrobial agent highlights its potential in tackling environmental and public health issues.

3.5.3 Sugarcane-based concrete alternatives for sustainable structures [53]

Residues from monocultures, including soybean production, corn cobs, straw, sunflower seeds, and cellulose, are commonly used for soil composting, animal feed, or energy conversion. This helps minimize waste and address the environmental consequences of agricultural activities.

Considering this perspective, the University of East London (UEL), in collaboration with Grimshaw Architects and manufacturer Tate & Lyle Sugar, has created an innovative construction material known as Sugarcrete™.

Figure 3.9 Sugarcrete™ [53].

The project's objective is to investigate sustainable building solutions through the utilization of recycled biological by-products from sugarcane. This approach effectively decreases carbon emissions in the construction industry while simultaneously emphasizing social and environmental sustainability throughout the production and application of these building materials (Figure 3.9).

The major innovation of Sugarcrete™ is its ability to challenge the prevailing belief that biomaterials have limited structural capabilities by developing a material that exhibits sufficient strength to support itself. Sugarcrete™ is produced through the amalgamation of bagasse with mineral binders. The final composite product has less weight than conventional brick and possesses only 15–20% of its carbon impact. By utilizing only 30% of the global bagasse production, Sugarcrete™ has the capacity to entirely substitute the conventional brick industry, leading to a reduction of 1.08 billion tons of CO2 emissions (equal to 3% of global carbon dioxide production).

REFERENCES

1. R. Shinde et al., "Management of crop residues with special reference to the on-farm utilization methods: A review," *Ind Crops Prod*, vol. 181, p. 114772, July 2022, doi: 10.1016/J.INDCROP.2022.114772
2. Y. G. T. G. S. Ballupete Nagaraju, M. Puttegowda, A. Verma, S. M. Rangappa, and S. Siengchin, "Biopolymer-based composites: An eco-friendly alternative from agricultural waste biomass," *J Comp Sci*, vol. 7, no. 6, p. 242, June 2023, doi: 10.3390/jcs7060242
3. M. V. Madurwar, R. V. Ralegaonkar, and S. A. Mandavgane, "Application of agro-waste for sustainable construction materials: A review," *Constr Build Mater*, vol. 38, pp. 872–878, January 2013, doi: 10.1016/J.CONBUILDMAT.2012.09.011

4. N. Mahawar, P. Goyal, S. Lakhiwal, and S. Jain, Agro waste: A new eco-friendly energy resource, *Int Res J Environ Sci*, vol. 4, no. 3, pp. 47–49, 2015, doi: 10.7251/AGSY1203393S
5. P. Rousu, P. Rousu, and J. Anttila, "Sustainable pulp production from agricultural waste," *Resour Conserv Recycl*, vol. 35, no. 1–2, pp. 85–103, April 2002, doi: 10.1016/S0921-3449(01)00124-0
6. S. Y. Lee et al., "Waste to bioenergy: A review on the recent conversion technologies," *BMC Energy*, vol. 1, no. 1, pp. 1–22, May 2019, doi: 10.1186/S42500-019-0004-7
7. R. Pratiwi, D. Rahayu, and M. I. Barliana, "Characterization of bioplastic from rice straw cellulose," *Res J Pharma Biol Chem Sci*, vol. 8, no. 1, pp. 217–221, 2017.
8. M. L. Sanyang, S. M. Sapuan, and M. Haron, "Effect of cocoa pod husk filler loading on tensile properties of cocoa pod husk/polylactic acid green biocomposite films," *AIP Conf Proc*, vol. 1891, no. 1, p. 20126, October 2017, doi: 10.1063/1.5005459/886824
9. D. Czajczyńska et al., "Potential of pyrolysis processes in the waste management sector," *Thermal Sci Eng Prog*, vol. 3, pp. 171–197, September 2017, doi: 10.1016/J.TSEP.2017.06.003
10. M. A. Morales, C. L. Atencio Martinez, A. Maranon, C. Hernandez, V. Michaud, and A. Porras, "Development and characterization of rice husk and recycled polypropylene composite filaments for 3D printing," *Polymers*, vol. 13, no. 7, p. 1067, March 2021, doi: 10.3390/POLYM13071067
11. S. Muthukrishnan, H. W. Kua, L. N. Yu, and J. K. H. Chung, "Fresh properties of cementitious materials containing rice husk ash for construction 3D printing," *J Mat Civil Eng*, vol. 32, no. 8, p. 04020195, August 2020, doi: 10.1061/(ASCE)MT.1943-5533.0003230
12. A. Yussuf, I. Massoumi, and A. Hassan, "Comparison of polylactic Acid/Kenaf and polylactic Acid/Rise husk composites: The influence of the natural fibers on the mechanical, thermal and biodegradability properties," *J Polym Environ*, vol. 18, no. 3, pp. 422–429, September 2010, doi: 10.1007/S10924-010-0185-0/FIGURES/5
13. C. Nyambo, A. K. Mohanty, and M. Misra, "Polylactide-based renewable green composites from agricultural residues and their hybrids," *Biomacromolecules*, vol. 11, no. 6, pp. 1654–1660, July 2010, doi: 10.1021/BM1003114
14. C. S. Wu, H. T. Liao, J. J. Jhang, J. T. Yeh, C. Y. Huang, and S. L. Wang, "Thermal properties and characterization of surface-treated RSF-reinforced polylactide composites," *Polym Bull*, vol. 70, no. 11, pp. 3221–3239, November 2013, doi: 10.1007/S00289-013-1018-9/FIGURES/8
15. E. Amir Kellersztein, and A. Dotan, "Grafting of wheat straw fibers with poly (ε-caprolactone) via ring opening polymerization for poly(lactic acid) reinforcement," *Polym Adv Technol*, vol. 27, no. 5, pp. 657–664, May 2016, doi: 10.1002/PAT.3736
16. N. H. M. Zubir, S. T. Sam, N. N. Zulkepli, and M. F. Omar, "The effect of rice straw particulate loading and polyethylene glycol as plasticizer on the properties of polylactic acid/polyhydroxybutyrate-valerate blends," *Polym Bull*, vol. 75, no. 1, pp. 61–76, January 2018, doi: 10.1007/S00289-017-2018-Y/FIGURES/12

17. M. Ghorbani Chaboki, J. Mohammadi-Rovshandeh, and F. Hemmati, "Poly(lactic acid)/thermoplasticized rice straw biocomposites: Effects of benzylated lignocellulosic filler and nanoclay," *Iranian Polym J*, vol. 28, no. 9, pp. 777–788, September 2019, doi: 10.1007/S13726-019-00743-1/FIGURES/9
18. C. H. Tsou et al., "Preparation and characterization of renewable composites from Polylactide and Rice husk for 3D printing applications," *J Polym Res*, vol. 26, no. 9, pp. 1–10, September 2019, doi: 10.1007/S10965-019-1882-6/TABLES/2
19. K. Chen, P. Li, X. Li, C. Liao, X. Li, and Y. Zuo, "Effect of silane coupling agent on compatibility interface and properties of wheat straw/polylactic acid composites," *Int J Biol Macromol*, vol. 182, pp. 2108–2116, July 2021, doi: 10.1016/J.IJBIOMAC.2021.05.207
20. W. Yu, L. Dong, W. Lei, Y. Zhou, Y. Pu, and X. Zhang, "Effects of rice straw powder (RSP) size and pretreatment on properties of FDM 3D-printed RSP/poly(lactic acid) biocomposites," *Molecules*, vol. 26, no. 11, p. 3234, May 2021, doi: 10.3390/MOLECULES26113234
21. V. Ganapathy et al., "Mechanical properties of cellulose-based multiscale composites: A review," *Polym Compos*, vol. 44, no. 2, pp. 734–756, February 2023, doi: 10.1002/PC.27175
22. S. B. Shenoy, and M. Vijaya Kini, "Effect of chemical treatments on hardness and toughness properties of *Grewia serrulata* reinforced polymer composites," 2019, doi: 10.26480/jmerd.04.2019.228.230
23. E. F. Cerqueira, C. A. R. P. Baptista, and D. R. Mulinari, "Mechanical behaviour of polypropylene reinforced sugarcane bagasse fibers composites," *Procedia Eng*, vol. 10, pp. 2046–2051, January 2011, doi: 10.1016/J.PROENG.2011.04.339
24. V. S. Candido, A. C. R. da Silva, N. T. Simonassi, F. S. da Luz, and S. N. Monteiro, "Toughness of polyester matrix composites reinforced with sugarcane bagasse fibers evaluated by Charpy impact tests," *J Mat Res Technol*, vol. 6, no. 4, pp. 334–338, October 2017, doi: 10.1016/J.JMRT.2017.06.001
25. N. A. Ramlee, M. Jawaid, E. S. Zainudin, and S. A. K. Yamani, "Tensile, physical and morphological properties of oil palm empty fruit bunch/sugarcane bagasse fibre reinforced phenolic hybrid composites," *J Mat Res Technol*, vol. 8, no. 4, pp. 3466–3474, July 2019, doi: 10.1016/J.JMRT.2019.06.016
26. E. F. Rodrigues, T. F. Maia, and D. R. Mulinari, "Tensile strength of polyester resin reinforced sugarcane bagasse fibers modified by estherification," 2011, doi: 10.1016/j.proeng.2011.04.387
27. V. Vilay, M. Mariatti, R. Mat Taib, and M. Todo, "Effect of fiber surface treatment and fiber loading on the properties of bagasse fiber–reinforced unsaturated polyester composites," *Compos Sci Technol*, vol. 68, no. 3–4, pp. 631–638, March 2008, doi: 10.1016/J.COMPSCITECH.2007.10.005
28. H. Y. Kordkheili, S. Hiziroglu, and M. Farsi, "Some of the physical and mechanical properties of cement composites manufactured from carbon nanotubes and bagasse fiber," *Mater Des*, vol. 33, no. 1, pp. 395–398, January 2012, doi: 10.1016/J.MATDES.2011.04.027
29. H. Luo et al., "Mechanical and thermo-mechanical behaviors of sizing-treated corn fiber/polylactide composites," *Polym Test*, vol. 39, pp. 45–52, 2014, doi: 10.1016/j.polymertesting.2014.07.014

30. V. Bridgwater, "Review of fast pyrolysis of biomass and product upgrading," *Biomass Bioenergy*, vol. 38, pp. 68–94, March 2012, doi: 10.1016/J.BIOMBIOE.2011.01.048
31. A. B. A. Mohammed, A. A. Borhana Omran, Z. Hasan, R. A. Ilyas, and S. M. Sapuan, "Wheat biocomposite extraction, structure, properties and characterization: A review," *Polymers*, vol. 13, no. 21, p. 3624, October 2021, doi: 10.3390/POLYM13213624
32. S. Panthapulakkal and M. Sain, "The use of wheat straw fibres as reinforcements in composites," *Biofiber Reinforc Comp Mater*, pp. 423–453, January 2015, doi: 10.1533/9781782421276.4.423
33. C. Nyambo, A. K. Mohanty, and M. Misra, "Effect of maleated compatibilizer on performance of PLA/wheat straw-based green composites," *Macromol Mater Eng*, vol. 296, no. 8, pp. 710–718, August 2011, doi: 10.1002/mame.201000403
34. S. Panthapulakkal, A. Zereshkian, and M. Sain, "Preparation and characterization of wheat straw fibers for reinforcing application in injection molded thermoplastic composites," *Bioresour Technol*, vol. 97, no. 2, pp. 265–272, January 2006, doi: 10.1016/j.biortech.2005.02.043
35. F. Le Digabel, N. Boquillon, P. Dole, B. Monties, and L. Averous, "Properties of thermoplastic composites based on wheat-straw lignocellulosic fillers," *J Appl Polym Sci*, vol. 93, no. 1, pp. 428–436, July 2004, doi: 10.1002/app.20426
36. S. Panthapulakkal and M. Sain, "Agro-residue reinforced high-density polyethylene composites: Fiber characterization and analysis of composite properties," *Compos Part A Appl Sci Manuf*, vol. 38, no. 6, pp. 1445–1454, June 2007, doi: 10.1016/j.compositesa.2007.01.015
37. X. R. Cui, Y. S. Wang, Y. Chen, H. Y. Mu, and H. H. Chen, "Understanding the digestibility of wheat starch- caffeic acid complexes prepared by hot-extrusion 3D printing technology," *Food Hydrocoll*, vol. 141, p. 108692, August 2023, doi: 10.1016/J.FOODHYD.2023.108692
38. V. Nagarajan, A. K. Mohanty, and M. Misra, "Sustainable green composites: Value addition to agricultural residues and perennial grasses," *ACS Sustain Chem Eng*, vol. 1, no. 3, pp. 325–333, March 2013, doi: 10.1021/SC300084Z/SUPPL_FILE/SC300084Z_SI_001.PDF
39. R. Zhu, H. Liu, and J. Zhang, "Compatibilizing effects of maleated poly(lactic acid) (PLA) on properties of PLA/soy protein composites," *Ind Eng Chem Res*, vol. 51, no. 22, pp. 7786–7792, June 2012, doi: 10.1021/IE300118X
40. L. Mohammed, M. N. M. Ansari, G. Pua, M. Jawaid, and M. S. Islam, "A review on natural fiber reinforced polymer composite and its applications," *Int J Polym Sci*, vol. 2015, 2015, doi: 10.1155/2015/243947
41. "Review of the History Properties and Application of Plant Fibres |Statistics of Plant Fibres|Fibre2fashion.com – Fibre2Fashion." Accessed: September 27, 2024. [Online]. Available: https://www.fibre2fashion.com/industry-article/3889/review-of-the-history-properties-and-application-of-plant-fibres
42. O. Awogbemi, and D. V. Von Kallon, "Valorization of agricultural wastes for biofuel applications," *Heliyon*, vol. 8, no. 10, p. e11117, October 2022, doi: 10.1016/J.HELIYON.2022.E11117
43. W. Tan et al., "Bamboo particle reinforced polypropylene composites made from different fractions of bamboo culm: Fiber characterization and analysis

of composite properties," *Polym Compos*, vol. 40, no. 12, pp. 4619–4628, December 2019, doi: 10.1002/PC.25329

44. X. Li, B. Lei, Z. Lin, L. Huang, S. Tan, and X. Cai, "The utilization of bamboo charcoal enhances wood plastic composites with excellent mechanical and thermal properties," *Mater Des*, vol. 53, pp. 419–424, January 2014, doi: 10.1016/J.MATDES.2013.07.028
45. S. C. Chin, K. F. Tee, F. S. Tong, H. R. Ong, and J. Gimbun, "Thermal and mechanical properties of bamboo fiber reinforced composites," *Mater Today Commun*, vol. 23, p. 100876, June 2020, doi: 10.1016/J.MTCOMM.2019.100876
46. A. Shahzad, "Hemp fiber and its composites – A review," *J Compos Mater*, vol. 46, no. 8, pp. 973–986, April 2012, doi: 10.1177/0021998311413623
47. S. Prasad, A. Singh, N. E. Korres, D. Rathore, S. Sevda, and D. Pant, "Sustainable utilization of crop residues for energy generation: A life cycle assessment (LCA) perspective," *Bioresour Technol*, vol. 303, p. 122964, May 2020, doi: 10.1016/J.BIORTECH.2020.122964
48. A. Dutta, A. Patra, K. K. Hazra, C. P. Nath, N. Kumar, and A. Rakshit, "A state of the art review in crop residue burning in India: Previous knowledge, present circumstances and future strategies," *Environ Chall*, vol. 8, p. 100581, August 2022, doi: 10.1016/J.ENVC.2022.100581
49. L. Chen et al., "3D-printed tripodal porous wood-mimetic cellulosic composite evaporator for salt-free water desalination," *Compos B Eng*, vol. 263, p. 110830, August 2023, doi: 10.1016/J.COMPOSITESB.2023.110830
50. A. Abdullah et al., "Potential for sustainable utilisation of agricultural residues for bioenergy production in Pakistan: An overview," *J Clean Prod*, vol. 287, p. 125047, March 2021, doi: 10.1016/J.JCLEPRO.2020.125047
51. "How this company is turning India's agri-waste into biodegradable plastic." Accessed: September 27, 2024. [Online]. Available: https://www.bizzbuzz.news/bizz-talk/how-this-company-is-turning-indias-agri-waste-into-biodegradable-plastic-1317348
52. H. S. Shahraki et al., "Papaya peel waste carbon dots/reduced graphene oxide nanocomposite: From photocatalytic decomposition of methylene blue to antimicrobial activity," *J Biores Bioprod*, vol. 8, no. 2, pp. 162–175, May 2023, doi: 10.1016/j.jobab.2023.01.009
53. "From Agro-Waste to Sustainable Structures: Concrete Alternatives Made from Sugarcane|ArchDaily." Accessed: September 27, 2024. [Online]. Available: https://www.archdaily.com/1001501/from-agro-waste-to-sustainable-structures-concrete-made-from-sugarcane.

Chapter 4

Advanced engineering applications of natural fiber composites for wind turbine blades

Oluwaseyi Omotayo Alabi, Amos O. Olaleye, Timileyin O. Akande, Oluwatoyin Joseph Gbadeyan, and Ayotunde O. Fasina

4.1 INTRODUCTION

Natural fiber composites (NFCs) have emerged as a promising alternative to traditional materials such as fiberglass and carbon fiber for use in wind turbine blades. The main advantages of NFCs are their lightweight, low cost, and biodegradability. Climate change, driven by human-caused emissions of greenhouse gases, poses a major threat to the world's ecosystems and population. Increased global temperatures are linked to more severe storms, droughts, and rising sea levels, all of which can have devastating effects on human settlements and wildlife. To address this urgent issue, world leaders gathered in Paris in 2015 and signed the Paris Agreement. This agreement aims to limit the rise in global temperatures to no more than 2°C above pre-industrial levels, with the ultimate goal of 1.5°C [1]. Achieving ambitious climate targets demands significant reductions in global greenhouse gas emissions. While multiple strategies can contribute to this endeavor, a particularly relevant approach is the adoption of renewable energy. Despite the ongoing development and deployment of diverse technologies, the most effective solution is likely to emerge from a synergistic blend of various renewable energy methods [2]. The technology associated with wind energy is notably fascinating due to its potential to compete with established energy sources in cost-effectiveness. Despite the widespread distribution of wind turbines globally, enhancing their efficiency requires overcoming significant scientific and technological challenges, thereby increasing the overall cost of energy production [1, 2].

Presently, ongoing research within the wind turbine industry aims to enhance the efficiency of these devices. This research has led to the creation of innovative wind turbine blade designs incorporating sophisticated material systems, as illustrated in Figure 4.1. The focus is on developing future blade structures that outperform existing thermoset technology in terms of lower density, extended lifespan, improved performance, ease of manufacturing, recyclability, and reduced cost. The adoption of alternative material systems is geared toward achieving a superior life cycle for the blades,

DOI: 10.1201/9781003527275-4

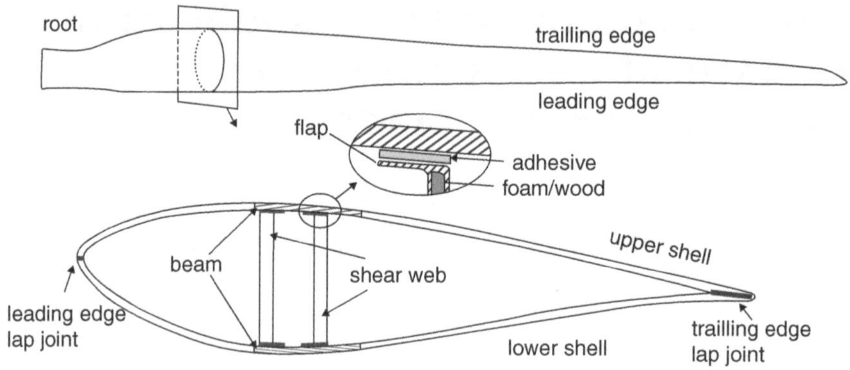

Figure 4.1 The schematic diagrams of a wind turbine blade and its cross-section.

mitigating gravity forces, optimizing aerodynamic performance, and minimizing material deterioration during operation [3].

One of the main industries using composite materials is wind energy. The nacelle and the blades are the two main components made of composite material [3, 4]. The protective housing covering the generator, drive train, and gearbox to shield them from environmental factors is known as the nacelle. Typically constructed from glass fiber composites, the nacelle prioritizes low weight, strength, and corrosion resistance. Despite its critical role, a wind turbine's most essential and cost-intensive component is arguably its blades [5]. These blades, often the priciest part of the turbine, significantly impact its lifespan and performance. Subjected to various load situations such as gravity forces, cyclic loads, dynamic wind forces, and more, the blades are pivotal to a wind turbine's functionality. International standards provided by organizations like ASTM, ISO, and others aim to meet the structural criteria for wind turbine blades, considering these diverse factors. In the structural design of the blades, emphasis must be placed on demonstrating a high strength-to-weight ratio of the materials [6]. Nevertheless, the weight of the blades is influenced by various factors, including material type, turbine geometry, and airfoil type. Achieving a robust blade surface quality necessitates not only suitable material selection but also advanced manufacturing processes. Deploying this is essential to remove unwanted outside blade weight from the surface, such as snow and mud. Thus, it is necessary to take into account all relevant elements, including wind turbine kinds, capacities, types of blade material, and manufacturing methods. Composites containing natural or synthetic fibers are used to create wind turbine blades [7–9]. Now, carbon and glass fibers are often used in big wind power plants, but aramid, aluminum alloy, and sometimes hybrid fibers can be used in medium- and small-scale blade designs to improve the performance of reinforced composites.

The use of E-glass/epoxy composites in wind blades can help to improve their structural performance and reduce the weight of the blades, leading to increased efficiency and lower energy costs [10]. E-glass/epoxy composites exhibit resistance to various environmental factors and offer a commendable strength-to-weight ratio. However, due to their non-biodegradable nature, they lack stiffness, possess limited resilience to high temperatures, and contribute significantly to negative environmental impact when disposed of as waste. An alternative option for wind blade production involves S-glass/epoxy composites, incorporating S-glass fibers [3, 11]. Compared to E-glass fibers, S-glass fibers exhibit superior strength and resilience and have a higher modulus and tensile strength. Despite these advantages, S-glass/epoxy composites are pricier, non-biodegradable, and more susceptible to abrasion under extremely high wind loads, despite their fatigue resistance and ability to withstand a diverse range of climatic conditions [12–16].

The literature review indicates the recent utilization of various composite materials in wind blade production. Composite materials predominantly composed of synthetic fibers are favored for their superior mechanical strength. However, the non-biodegradable nature of these synthetic composites makes them environmentally unfriendly, which has motivated continuing research into viable biodegradable alternatives for wind blade manufacture. The potential of natural fibers, such as jute, hemp, bamboo, sisal, and coir, is therefore the subject of several research. This present study stands out for its innovative approach, investigating the utilization of flax fiber-based wind blades specifically designed for micro-scale wind turbine applications.

4.2 ADVANCED MATERIAL SYSTEM

The selection of materials, particularly polymers and fibers employed in constructing systems like wind turbine blades, has gained increasing economic significance in the wind energy sector. The cost and length of blades have risen to enhance turbine efficiency, directly linked to blade size. However, the mechanical characteristics of existing materials impose limits on the maximum size of wind turbine blades. Considering a material's density becomes crucial as the weight resulting from the size and density of the blades subjects the system to stress beyond the material's load-bearing capacity, leading to potential collapse. The solution to this problem is an emphasis on easily worked-with, reasonably priced materials that do not sacrifice the effectiveness and dependability of wind turbine blades [5, 17]. The current blade material system, while mechanically robust, poses environmental concerns due to its non-biodegradable nature. The development of more affordable, easily processed materials that maintain the functionality and reliability of wind turbine blades is imperative. The non-biodegradable composition of the current generation of wind turbine blades necessitates an alternative end-of-life solution, as landfill disposal is no longer acceptable due to the environmental risks posed by the blades' organic content. The study suggests a more effective approach would be to establish a recyclable material system, wherein fibers are collected and reintegrated into blade production through a closed-loop recycling system. Challenges in recycling synthetic materials, particularly carbon fiber, underscore the need for further research to lower recycling costs [18]. Achieving a material for wind turbine blades that balances cost-effectiveness, performance, and weight reduction poses a challenge for blade designers. The subsequent sections delve into the advantages and disadvantages of an advanced material system with high specific mechanical properties and eco-friendly, biodegradable qualities.

Numerous studies have showcased the potential of various materials for the production of wind turbine blades, reporting exceptional mechanical performance for thermoset and thermoplastic composite materials, respectively [6]. Various research endeavors have highlighted the effectiveness of diverse materials in wind turbine blade construction, particularly emphasizing the superior mechanical attributes of thermoset and thermoplastic composite materials, as evidenced by reports from Brondsted et al. (2005) and Gardiner (2012) [6]. Fibers utilized in wind turbine blades can take various forms, ranging from structurally elongated crystals to fragmented, short, or continuous fibers. Glass fibers, carbon fibers, and other types have been widely employed in the creation of wind turbine blades. The economic appeal of hybridizing carbon and glass fibers lies in their respective high performance and outstanding processability. Despite their synthetic nature, both glass and carbon fibers, which serve as reinforcing materials, face limitations in terms of post-service life recycling or biodegradation. Additionally, their potential adverse effects during processing pose significant drawbacks. In contrast to the vast amount of studies that have been done on natural fiber-reinforced composites, monolithic materials like glass and carbon fibers have not gotten as much attention. The alteration of composite material ingredients, including fibers and polymers, during the blade development process, is a common practice to achieve improved performance or desired qualities for wind turbines.

Additionally, being linear elastic materials, carbon fiber composites struggle to distribute stress concentrations and, like other ductile materials, have significant notch sensitivity. When a low-elongation layer fails, the material's vertical stress significantly decreases, which is unsuitable for the development of wind turbine blades. This is where pseudo ductility becomes crucial, particularly in composites with hybrid fiber reinforcement. Swolfs (2019) proposed the use of carbon fiber, with its higher density compared to the more common glass fiber, as a reinforcement material. However, using carbon fiber has drawbacks as well. These include brittleness, which increases its susceptibility to impact damage, sensitivity to misalignment due to anisotropy and increased stiffness, and difficulties in impregnation due to its tiny diameter and accompanying smaller interface spacing [8]. Although these drawbacks become less significant during operation, careful consideration is essential during transportation and installation. Good fatigue strength, decreased weight, high strength, and good stiffness are material qualities that are examined to solve issues including cycle loading, gravity forces, wind forces, and stability. Their research indicated that reinforcing composites with nanomaterials, specifically Carbon Nanotube Fibers (CNTs), yielded more favorable mechanical properties [19]. Because of the remarkable qualities of cylindrical carbon molecules, carbon nanotubes (CNTs), which belong to the carbon allotropy and have a nanostructure with an aspect ratio of more than 1,000,000, were found to be a better choice for wind turbine blades. According to Pradeep et al. (2019), the manufacturing sector has been compelled to explore environmentally friendly materials, such as thermoplastics and natural bio-composites, owing to technological advancements. Carbon nanotube-reinforced materials are preferable for composite structures due to the crucial requirements of low density and high strength [19, 20]. Moreover, hybridizing these carbon nanotubes with natural fibers could lead to the creation of composite materials for wind turbine blades that are not only lighter and more durable but also environmentally beneficial.

4.3 MECHANICAL CHARACTERISTICS OF NATURAL FIBER

4.3.1 Natural fibers

Due to its favorable attributes, including cost-effectiveness, relatively good mechanical properties, high specific strength, non-abrasiveness, biodegradability, and environmental friendliness, natural fiber has garnered popularity as a reinforcement for fiber-reinforced polymers (FRPs). These fibers exist in diverse forms and are sourced from minerals, plants, and animals. The categorization of natural fibers is illustrated in Figure 4.2 [9].

Bamboo is a material with tremendous potential for reinforcement in FRPs. Its high lignin concentration and 60% cellulose content allow bamboo fibers to function as reinforcement in a variety of polymer matrices. In a study by J. W. Holmes et al. (2018), the mechanical properties of glass, birch, and bamboo laminates were compared, revealing that the properties of the latter two were comparable to glass-reinforced polymer laminates, with bamboo exhibiting greater strength, fatigue life, and fracture resistance [21]. According to different research by Honey Banga et al. (2019), the particulate filled bamboo-reinforced polymer absorbs less water than unfilled bamboo-reinforced polymer, and 30% of bamboo is the ideal amount of bamboo to include in the laminate for mechanical qualities [21, 22]. In a comprehensive analysis by [13] comparing the performance of bamboo turbine blades to glass fiber turbine blades, the characteristics of bamboo

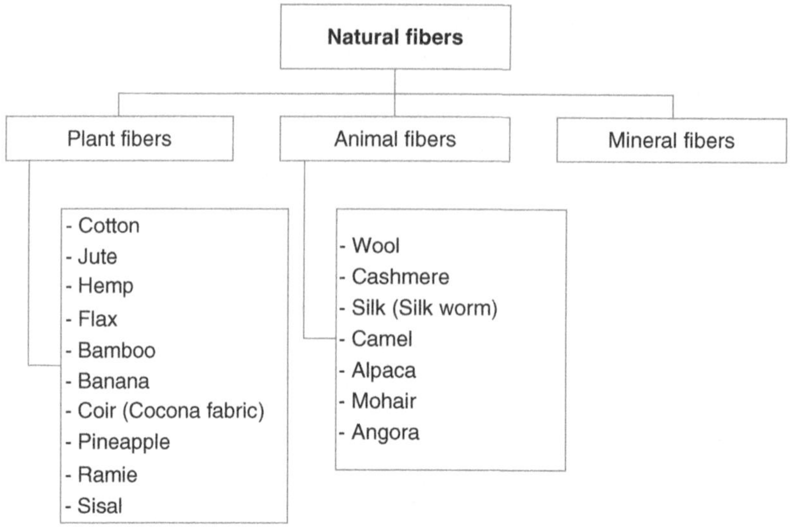

Figure 4.2 Natural fiber.

material were explored, and experiments with life cycle analysis were conducted. The results demonstrated that bamboo material met the requirements for wind turbine blades, further emphasizing its suitability for this application.

Abraham and Temesgen (2022) [5], delved into the mechanical properties of coir fiber composites, exploring tensile, impact, shear, flexural, and compression strength. Their findings revealed that, while the qualities of the composite are comparable to wood, they fall short of the superior properties exhibited by glass fiber composites. Additionally, the environmental impact on coir fiber composites was observed to reduce the material's mechanical properties with increasing weathering times. Nottingham University in the United Kingdom successfully crafted a 3.5-meter-long wind turbine blade using 600GSM flax biaxial ±45, meeting the IEC61400 standard flax fiber, when blended with a polymer mixture, exhibits high specific strength and superior mechanical qualities compared to other natural fibers [3]. However, Abdul Nasir et al. (2018) explored the impact of machining on flax fiber, highlighting a drawback of brittle fracture leading to delamination due to the high cellulose concentration of flax fiber [23]. This drawback underscores a challenge when employing NFCs in wind turbine blades. To mitigate issues like moisture absorption, sizing, and other reinforcement, fiber treatments become essential. Sizing, in particular, facilitates effective stress transfer between the fiber and resin by enhancing the fiber surface's wettability, enabling a better connection with the polymer [21].

Since natural fibers are inexpensive, readily available, biodegradable, eco-friendly, non-abrasive, and have excellent mechanical properties, their use as reinforcement for wind turbine blades has increased dramatically. Abaka, henequen, pineapple, banana, seed fibers (kapok, cotton, doum palm nuts), stalk fibers (rice and bamboo), fruit fibers (coir and tamarind), stem fibers (sisal and crown), and bast fibers (hemp, jute, flax, kenaf, roselle, ramie, rattan) are a few examples of commonly used natural fibers. Additionally, particulates like groundnut shells, seashells, and jackfruit are also incorporated. These natural fibers are typically sourced from various parts of plants, showcasing excellent mechanical qualities [1]. An analysis of various studies (as summarized in Table 4.1) reveals that natural fibers, on average, possess tensile strengths exceeding 500 MPa. Notably, some natural fibers, including cotton, flax, kenaf, and wool, exhibit exceptional tensile strengths of 600, 1450, 890, and 900 MPa, respectively. This highlights the considerable comparability of their mechanical properties to some synthetic fibers. Athijayamani et al. (2009) reported specific characteristics of these natural fibers, noting that flax fiber exhibited a tensile behavior reaching 1400 MPa, with an average density of 1400 kg/m^3 [6].

The exploration of wind energy resources has extended to extremely hostile locations, prompting the investigation of high-performing and environmentally friendly materials, such as those derived from bio-nanocomposites. Kausar's (2020) study, focusing on this specific category of materials for such applications, forecasts a future emphasis on greener and more environmentally friendly materials. These materials are poised to compete based on crucial attributes like low weight, high efficiency, and easy obtainability, ensuring affordability, sustainability, and lightweight characteristics [15].

Table 4.1 Certain natural fibers' mechanical properties [9]

Fiber	*Density (Kg/m^3)*	*Tensile strength (MPa)*	*Author*
Cotton	1500–1600	300–607	[6]
Coir	1200	180–230	[9]
Flax	1400–1500	365–1450	[22] [9]
Hemp	1480	600–890	[9]
Jute	1300	400–790	[9]
Kenaf	1400	890	[13]
Ramie	1500	230–951	[21] [1]
Sisal	1330–1500	410–690	[6]
Wool	1500	900	[13]
Bamboo	1500	400	[21]
Banana	1450	376	[9]
Cashmere	1400	610	[22]
Camel	1300	402	[1]
Pineapple	1700	700	[6] [13]
Angora	1800	800	[1]

The unique qualities of environmentally friendly reinforcing and sustainable polymers distinguish green polymeric nanocomposites. These environmentally friendly nanocomposites, vital in the energy production sector, incorporate nanofillers for reinforcement. Initially, the use of petroleum-based polymers posed a serious threat to the ecosystem. However, the adoption of ecologically friendly nanofillers has proven to be a beneficial substitute, mitigating the environmental impact. Integrating environmentally friendly cellulose nanofibers and inorganic nanofillers into specific synthetic polymers enhances their performance and paves the way for further research (Kumar, 2019) [25]. The development of materials like hybrid plastics, combining synthetic and natural polymers, and bio-nanocomposites has the potential to yield high-performance materials. A thorough analysis is necessary to weigh the trade-off between structural and mechanical qualities in comparison to synthetic materials, even though producing materials solely from natural sources (also known as green composites or nanocomposites) might save costs and improve environmental friendliness [26]. Predicted physical features, ease of production, and processing methods all contribute to the success of these eco-friendly (green) materials, with the increasing size of wind turbine blades, both current and future, adding complexity to the balancing act of these properties.

4.3.2 Natural fiber composites

While manufacturing materials entirely from natural sources (also called green composites or nanocomposites) may result in cost savings and increased environmental friendliness, a careful examination is required to assess the trade-off between structural and mechanical qualities in comparison to synthetic materials. Ng et al. (2019) discovered that the pineapple/polypropylene composite has a tensile strength of 17.07 MPa, and flexural strengths and moduli were 2.75 and 45.25 MPa [27]. Yallew et al. (2020) contributed research indicating tensile strengths of 56.51 MPa for hemp/polypropylene composites, 41.48 MPa for jute/polypropylene composites, and 75.84 MPa for sisal/polypropylene composites. These findings point to a bright future for the use of natural fibers as reinforcing materials in wind turbine blade assembly [28]. Figure 4.3 illustrates the mechanical properties of select NFCs, showcasing their overall commendable performance, with jute standing out for its exceptional characteristics [9]. It is worth noting that the mechanical strength of the reinforced material surpasses that of the virgin matrix. Notably, epoxy and polyester-based composites outperform other polymers in terms of mechanical properties. The strength of the bond or adhesion between the fiber and the polymer matrix is crucial, significantly influencing the mechanical characteristics of natural fiber and polymer composites (Table 4.2).

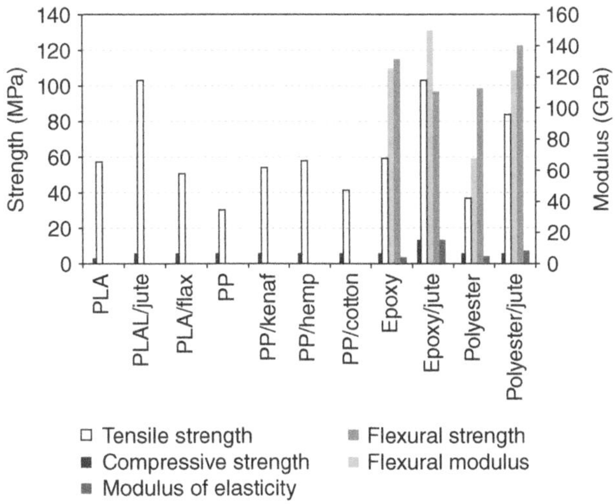

Figure 4.3 Mechanical properties of some selected natural fiber composites.

Table 4.2 Certain natural fiber composites' mechanical properties

Composite	*Tensile strength (MPa)*	*Flexural strength (MPa)*	*References*
Sisal/epoxy	95–330	315	[19]
Sisal/PE	70.1	100	[18]
Sisal/B/E	71	80	[14]
Flax/epoxy	270	308	[16]
Flax/VE	120–239	130	[9]
Flax/UP	288	195	[18]
Flax/PP	39–295	69–150	[5]
Hemp/epoxy	149	170	[9] [8]
Hemp/PP		130	[14]
Hemp/PLA	59–78	99–132	[9]
Kenaf/PLA	79–228	130–243	[18]
Kenaf/PHB	69.9	99.85	[8]
Kenaf/PP	53	64	[3]
Jute/UP	52	110	[19]
Jute/PP	80	109	[14]
PALF/UP	51	84	[16]
Banana/epoxy	35	130	[19]
Rattan	14	129	[5]

4.4 NATURAL FIBER HYBRID COMPOSITES

According to Swolfs (2018), fiber-hybrid composites are composites that combine different types of fiber into a matrix to provide greater design flexibility. They have made significant progress in the last few years, but more study is still required to fully understand these hybrid fiber composites, especially the ones that have faux ductility [9]. While many composites exhibit excellent strength and stiffness, the challenge lies in their lack of ductility, especially crucial for applications like wind turbine blades. Fiber hybridization has emerged as a solution to address both strength and ductility issues. Ductility is a crucial quality for most engineering applications, influencing the failure pattern of structures made from these materials. Pseudocontility, which is defined as the strain difference between the ultimate failure strain and the extrapolated linear elastic area at the same stress level, is another solution proposed by other researchers for non-ductile fiber hybrid composites [29]. The concept of pseudo ductility, as presented by Swolfs, holds the potential to alleviate significant drawbacks associated with composite materials in the realm of wind turbine development. When compared to more ductile alternatives, these materials frequently show little to no warning signals before breakdown, raising safety issues and resulting in less-than-ideal design. The combination of synthetic and natural fibers has been a key focus in many hybrid composites. This approach leverages the enhanced mechanical properties of synthetic fibers [30]. For instance, Akil et al. (2014) combined high modulus jute with glass fiber, resulting in a composite with significantly improved tensile and flexural capabilities. The presence of glass fiber played a crucial role in enhancing mechanical properties while reducing water absorption characteristics [31].

It is equally noteworthy that two natural fibers can hybridize to create a composite material that offers a chance to improve properties and possibly save expenses. This approach has garnered interest in the realm of biodegradable composites, where both the polymers (matrix) and fibers (reinforcement) are biodegradable [32]. While synthetic-natural hybrid composites outperform natural hybrids, offering greater potential for wind turbine blade development, there remains significant potential in exploring ways to enhance the performance of natural hybrid composites [29, 32]. The low cost and unique qualities of these composites, including hybrid nanocomposites, should drive further research to develop materials that are both environmentally friendly and competitive in mechanical attributes compared to pure synthetic composites or hybrids combining synthetic and natural materials [33, 34].

4.5 MECHANICAL RECYCLING METHODS FOR COMPOSITE WIND TURBINE BLADES

Currently, the predominant recycling method involves mechanical grinding, where composite materials, comprising fibers and a polymer matrix, are ground into particles and incorporated as fillers in products like asphalt or cement. Despite its simplicity and cost-effectiveness, this method has drawbacks as it doesn't recover fiberglass or epoxy resin for future use [35]. A more recent recycling technique, pyrolysis, addresses this limitation by heating composite materials in an oxygen-free environment, breaking them down into their constituent elements. Through pyrolysis, both fiberglass and epoxy resin can be extracted and repurposed to manufacture new products. Composite materials used in wind turbine blades typically consist of fibers such as glass, carbon, natural, or hybrid fibers (combining two or more fibers of different types) mixed with a polymer matrix, which could be thermoplastic or thermosetting [14]. Thermosetting polymer matrices pose challenges for recycling due to the crosslinking phenomena in their molecular chains [14, 36]. Currently, the wind industry primarily employs two recycling techniques: the mechanical approach and the thermal approach [35].

Retiring wind turbine blades undergo initial slicing into small pieces onsite to facilitate transportation to the recycling plant for both recycling methods. In the mechanical approach, the next step involves crushing the blade pieces with a mechanical hammer to produce grains of varying sizes, such as coarse grains measuring 10 mm and fine grains at 50 μm [37]. These recycled materials can be utilized in diverse projects, including civil engineering. On the other hand, the thermal approach requires placing the various cut parts into a rotary kiln incinerator after the cutting process and subjecting them to pyrolysis at 500°C. The resulting materials find applications in various fields, such as civil engineering. The fiber residues generated through this process can be repurposed for structural purposes, such as in concrete for building and road construction [38]. They may also be integrated into other composite structures serving as structural elements under less stress.

In practice, research suggests that glass fibers are more commonly employed in wind turbine blade production due to their lower cost. However, the recyclability issue with thermosetting polymer matrices like polyester and epoxy diminishes their appeal among blade manufacturers. In contrast, the thermoplastic polymer matrix emerges as a highly attractive option for future wind turbine blade production, offering easy recyclability post-use and promoting sustainability. An example of this is the innovative anionic polyamide PA-6 thermoplastic matrix designed for the wind blade sector [39]. Blade designers with environmental consciousness must strategically consider the impact of their products on the environment and devise novel approaches to enhance competitiveness in the global composite market. These approaches may include implementing precise safety standards, establishing clear legal frameworks, and adhering to transparent regulations [40]. The eco-design concept introduces a fresh perspective, incorporating new criteria for innovative blades alongside traditional design standards. These additional criteria encompass waste management, VOC emission reduction, optimized energy consumption, waste recovery systems, and specialized recycling methods for retired blades. An eco-design key performance indicator (KPI) can be developed based on probability theory principles to address these requirements. This KPI empowers blade designers to assess the performance level at each design step using environmentally friendly criteria. For more in-depth information on this topic, readers are encouraged to refer to previous research conducted by the author and published elsewhere [40, 41].

FRPs are generally milled, crushed, or shredded in the context of mechanical recycling. Fiber-rich powdered goods and fibrous components with a high resin content are the next categories for the remaining materials [9, 39]. These retrieved elements can be subsequently integrated into asphalts and cement or used as reinforcement or filler by reimpregnating them with fresh resin [42]. Particle grades are mostly used as filler, whereas fibrous components which have a greater aspect ratio are used as reinforcement [43]. In Sheet Molding Compound (SMC), for example, research has shown that recycled composite particles may replace up to 88 weight percent of the calcium carbonate without appreciably changing the final product's tensile and flexural properties [44]. When utilizing recycled composites in SMC and Bulk Molding Compound (BMC), there are a few technological issues to take into account, such as the uncured resin's increased viscosity and the delayed crosslinking reaction caused by the material's decreased heat conductivity [45]. In another example, the replacement of $CaCO_3$ filler with finely powdered (size less than 20 mm) composite waste recovered from its production stream led to notable improvements in damping and stress at break properties [46]. Recycled composites can also be used as filler or reinforcement in thermoplastic matrices (like PP), which will reduce the Charpy impact strength and significantly increase stress at break [47]. Many examples of how ground and sieved composite recycles affect the mechanical characteristics of different polymer matrices are given in Goodship's book. As an alternative, concrete can be strengthened in both the compressive and flexural directions by adding mechanically recycled composites that have been broken up and mixed in with a significant level of ductility [48].

4.5.1 Recycling of composite materials for wind blades

Metals (steel, aluminum, and copper), epoxy resin, and reinforcing fibers (GFs, CFs) make up the majority of a wind turbine's construction. Because of their more homogenous composition and higher economic worth, metals are often recycled at higher rates than glass and carbon fibers. An overview of the material quantities related to wind turbines (WTs) and the disposal strategies used at the end of their useful lives is shown in Figure 4.4 [49]. Reports from various researchers suggest that epoxy/GF composites (with a Vf of about 50%) are commonly used in the production of wind blades for plants with a nominal power. This choice is driven by the strong mechanical and chemical resistance of epoxy resin and the wide availability and reasonable pricing of GFs [50]. Regardless of the manufacturing method,

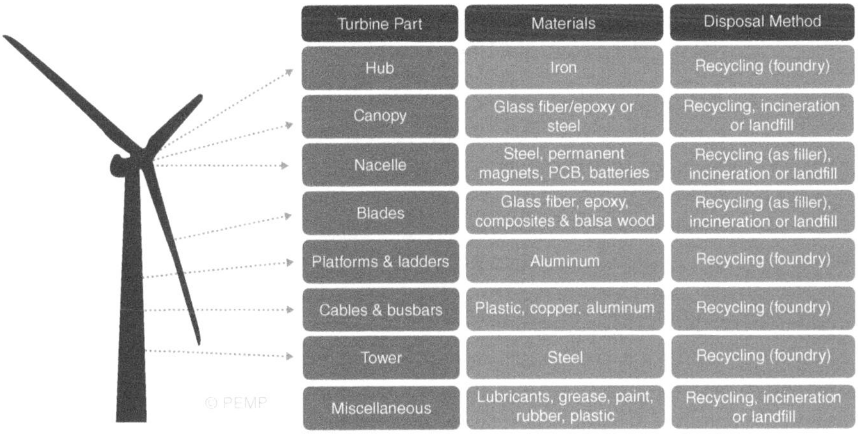

Figure 4.4 Overview of the material volumes associated with wind turbines (WTs) and their end-of-life disposal.

the total waste generated in producing a single blade is typically around 10%. Furthermore, consumables like peel ply, breeding cloth, vacuum bags, and vacuum tape that are utilized in the infusion process must be taken into account. This means that about 120 kg of consumables and about 540 kgs of manufacturing waste must be disposed of for a 5.4 t blade. Consequently, the composite materials of the blades account for about 18.6 t for a wind turbine with a nominal output of 1.65 MW.

Given the escalating use of wind energy, it is anticipated that in the next two decades, waste composites originating from WTs will exceed 106 t. Projections suggest that global rotor blade recycling will reach approximately 225,000 t annually by 2034 [51, 52]. As such, it is essential to conduct a thorough analysis of the composite materials end-of-life situation. Although landfilling is no longer an economically viable alternative because of the massive quantities required, the analysis provided in the Goodship book suggests that incineration with energy recovery might serve as a near-to-medium-term solution. Instead, the most viable approach going forward may involve recovering composites from wind blades and recycling them using various methods [53]. Still, the dearth of studies on life cycle assessment (LCA) studies for WTs and the small number of industrial trials demonstrate how important it is to do extensive research to solve the compound waste problem. Establishing if there is a market for the recovered materials is the most important economic consideration when examining the issue of composite recycling [54]. In practice, companies that handle composite waste have significant obstacles to transportation expenses and economies of scale, which can only be overcome by reliable feedstock availability.

Several instances of recycled composite wind blade constructions are given in the following paragraphs.

4.5.2 Innovative solutions for composite blades with improved recyclability

Based on the data in the sections before it, it is clear that recycling glass fiber (GF) and carbon fiber (CF) reinforced thermoset composites effectively presents obstacles and high recycling expenses. As a result, a great deal of research has been done on novel recyclable composite materials to use in the wind industry. The three main categories of FRPs that now demonstrate improved recyclability are thermoplastic composites, NFCs, and modified thermosetting composites [55].

There has been a recent push to incorporate active covalent connections in epoxy systems, given that thermosetting polymers are in general insoluble and infusible [56, 57]. Covalent bonds can break and recombine in response to environmental stimuli including light, heat, and radiation. Stated differently, covalent adaptive networks (CANs) allow shape change or stress reduction processes to be activated, thermosetting polymers to be recycled or reprocessed, and materials to repair or self-heal. The second possibility involves treating and reshaping an epoxy matrix like that of thermoplastics. If the dynamic macromolecular structure of CANs is to be attained via bond exchange or by equilibrium changes leading to reversible depolymerization, then CANs may be classified into two categories.

Exchangeable bonds in CAN systems belonging to the first category split and reassemble, maintaining the total number of bonds, the crosslinked structure, and the matrix's original stiffness [58]. The crosslinked structure, initial matrix stiffness, and total number of links are all preserved during this operation. Most recently, there has been interest in CANs based on moieties that are capable of addition-fragmentation chain transfer (AFT), like allyl sulfides or trithiocarbonates. AFT-based CANs involve the thiyl radical-mediated addition fragmentation of an allyl sulfide. Under these conditions, the action of a photo-initiator based on radicals facilitates the generation of radical species and, as a result, bond exchange reactions when exposed to light. By rearranging photo-responsive trithiocarbonate (TTC) units, Amamoto et al. (2013), for example, showed that photoinduced recurrent self-healing of covalently crosslinked structures is possible. Furthermore, utilizing flexible polyimine networks and woven CF sheets, Taynton et al. (2014) described how to create Carbon Fiber Reinforced Polymers (CFRPs) that are purely recyclable [59, 60]. Through dynamic covalent bond exchange with an excess of a monomer, the researchers were able to dissolve the polyimine binder and recover both the whole length of the fiber and the matrix. Using flexible, covalently crosslinked polyimines that dynamically swap covalent bonds in response to chemical or physical stimuli, this novel method enabled closed-loop recycling of these composites.

Diels-Alder adducts cause polymerization-depolymerization processes in thermoreversible bond topologies, which fall under the second class of CAN systems [61, 62]. This class studies the equilibrium reaction between a dienophile based on maleimide and a diene based on furan, identifying the production of crosslinked structures at low temperatures and the depolymerization that occurs upon heating during the Diels-Alder adduct process. Although there are very few active covalent bonds and very strict remolding requirements, this thermoresponsive CAN-based method may prove to be a good substitute for recycling FRPs [63, 64]. This thermoresponsive CAN-based approach could work well as a replacement for recycling FRPs, despite the high remolding needs and scarcity of active covalent bonds. However, further study is needed to progress the creation of these systems and increase the range of industrial environments in which they may be used.

4.6 CONCLUSION

The exploration of NFCs for wind turbine blades represents a promising avenue in the quest for sustainable and environmentally friendly energy solutions. The research and advancements in this field have unveiled a spectrum of possibilities that can significantly contribute to the optimization and performance enhancement of wind energy systems. Several key takeaways emerge from the examination of advanced engineering applications of NFCs in this context. The integration of NFCs into wind turbine blades aligns with the global pursuit of environmentally sustainable practices. These composites, derived from renewable resources, present an eco friendly alternative to traditional synthetic fibers. The reduced environmental impact during production and end-of-life disposal addresses concerns about the ecological footprint of wind energy technologies. Studies showcased in this exploration underscore the commendable mechanical performance of NFCs. The combination of specific natural fibers, such as bamboo, flax, and coir, with suitable matrices, has demonstrated competitive tensile, flexural, and impact strengths. The tailored mechanical properties of these composites make them viable candidates for the demanding conditions encountered by wind turbine blades.

A tactical technique that incorporates both natural and synthetic fibers is the hybridization of NFCs. This fusion capitalizes on the high performance of synthetic fibers like glass and carbon while harnessing the environmental benefits of natural fibers. The synergy achieved through hybridization enhances the overall mechanical and processing characteristics of the composites, rendering them economically attractive. While the sustainability aspect of NFCs is evident, challenges in recycling processes have been acknowledged. Innovations such as incorporating nanomaterials, developing biodegradable matrices, and exploring recycling methods offer potential solutions. These avenues represent ongoing efforts to address the life cycle considerations and ensure the responsible management of materials in wind turbine blade applications. As the exploration of NFCs for wind turbine blades continues, the future holds exciting prospects. Further research into refining manufacturing processes, optimizing material combinations, and addressing recycling challenges is imperative. Collaboration between academia, industry, and policymakers is vital to propel these advancements into mainstream wind energy applications. Continued research and innovation in this realm promise to reshape the dynamics of wind energy and contribute significantly to a cleaner and more sustainable energy future.

REFERENCES

1 Thomas, L. and Ramachandra, M. 2018. ScienceDirect Advanced Materials for Wind Turbine Blade: A Review, *Mater. Today Proc.*, vol. 5, no. 1, pp. 2635–2640, doi: 10.1016/j.matpr.2018.01.043

2 Meenakshi, C. M., Krishnamoorthy, A. and Reddy, C. V. 2019. Advancement in Engineering, Science & Technology, *J. Mech. Cont. & Math. Sci.*, no. 2, pp. 707–719, doi: 10.26782/jmcms.spl.2019.08.00086

3 Kalagi, G. R., Patil, R. and Nayak, N. 2018. ScienceDirect Experimental Study on Mechanical Properties of Natural Fiber Reinforced Polymer Composite Materials for Wind Turbine Blades, *Mater. Today Proc.*, vol. 5, no. 1, pp. 2588–2596, doi: 10.1016/j.matpr.2017.11.043

4 Robertson, I. D. 2018. Rapid Energy-Efficient Manufacturing of Polymers and Composites via Frontal Polymerization Composite, *Springer Nature, Letter*, vol. 5, no. 1, pp. 88–96.

5 Abraham, T. and Temesgen, M. 2022. Heliyon Experimental Characterizations of Hybrid Natural Fiber-Reinforced Composite for Wind Turbine Blades, vol. 8, January, doi: 10.1016/j.heliyon.2022.e09092

6 Pradeep, A. V., Prasad, S. V. S., Suryam, L. V. and Kumari, P. P. 2019. Proceedings a Comprehensive Review on Contemporary Materials Used for Blades of Wind Turbine, *Mater. Today Proc.*, pp. 1–4, doi: 10.1016/j.matpr.2019.07.732

7 Premkumar, V., Krishnamurty, S., Wileden, J. C. and Grosse, I. R. 2014 Advanced Engineering Informatics A Semantic Knowledge Management System for Laminated Composites, *Adv. Eng. Inform.*, vol. 28, no. 1, pp.91–101, doi: 10.1016/j.aei.2013.12.004

8 Sai, S., Reddy, P., Suresh, R., Hanamantraygouda, M. B. and Shivakumar, B. P. 2021. Proceedings Use of Composite Materials and Hybrid Composites in Wind Turbine Blades, *Mater. Today Proc.*, vol. 46, pp. 2827–2830, doi: 10.1016/j.matpr.2021.02.745

9 Alabi, O. O., Adeyi, T. A., Olagunju, G. O., Ogoh, O. A. and Gbadeyan, O. J. 2024. Investigation and Exploring the Microstructural and Thermal Properties of Pseudostem Juice of Banana and Cassava Juice as a Quenching Media, *Waste Biomass Valoriz. (Springer)*, vol. 2024, doi: 10.1007/s12649-024-02703-0

10 Batu, T., Lemu, H. G. and Sirhabizuh, B. 2020. Study of the Performance of Natural Fiber Reinforced Composites for Wind Turbine Blade Applications, *Elsevier Mater. Today, Proc.*, vol. 14, no. 2, pp. 67–75.

11 Debnath, K., Singh, I., Dvivedi, A. and Kumar, P. 2013. Recent Advances in Composite Materials for Wind Turbine Blades. *Adv. Mater. Sci. Appl.*, World Academic Publishing, Book Chapter, vol. 5, no. 3, pp. 6–229.

12 Dorigato, A. 2021. Advanced Industrial and Engineering Polymer Research Recycling of Thermosetting Composites for Wind Blade Application, *Adv. Ind. Eng. Polym. Res.*, vol. 4, no. 2, pp. 116–132, doi: 10.1016/j.aiepr.2021.02.002

13 Boria, S., Santulli, C., Raponi, E., Sarasini, F. and Tirillò, J. 2019. Evaluation of a New Green Composite Solution for Wind Turbine Blades, *Multiscale Multidiscip. Model. Exp. Des.*, no. April, doi: 10.1007/s41939-019-00043-4
14 Muthivhi, J., Linganiso, L. Z., Deenadayalu, N. and Alabi, O. O. 2024. Biogas Production and Techno-Economic Feasibility Studies of Setting Up Household Biogas Technology in Africa: A Critical Review, no. July, pp. 1–19, doi: 10.1002/ese3.1887
15 Karthikeyan, R. B., Suthakar, T. and Barhate, S. 2018. Materials, Innovations and Future Research Opportunities on Wind Turbine Blades Insight Review, pp. 1–10, doi: 10.1002/ep.13046
16 Amroune, S. 2021. Manufacturing of Rapid Prototypes of Mechanical Parts Using Reverse Engineering and 3D Printing, *J. Serbian Soc. Comput. Mech.*, vol. 15, no. 1, pp. 167–176, doi: 10.24874/jsscm.2021.15.01.11
17 Park, H. 2015. A Study on Structural Design and Analysis of Small Wind Turbine Blade with Natural Fibre (flax) Composite, no. June, pp. 37–41, doi: 10.1080/09243046.2015.1052186
18 Cao, D., Malakooti, S., Kulkarni, V. N., Ren, Y. and Lu, H. 2020. Nanoindentation Measurement of Core – Skin Interphase Viscoelastic Properties in a Sandwich Glass Composite, *Mech. Time-Depend. Mater.*, doi: 10.1007/s11043-020-09448-y
19 Speed, L., Turbine, W., Kasagepongsan, C. and Suchat, S. 2021. Novel Engineered Materials: Epoxy Resin Nanocomposite Reinforced with Modified Epoxidized Natural Rubber, *Polymers*, vol. 13, p. 2761. doi: 10.3390/polym13162761
20 Singh G. P. and Mangal, R. 2023. Proceedings a Comprehensive Review of Natural Fiber Reinforced Composite and their Modern Application, *Mater. Today Proc.*, doi: 10.1016/j.matpr.2023.03.745
21 Alabi, O. O., Gbadeyan, O. J., Towoju, O. A. and Deenadayalu, N. 2024. Enhancing Sustainable Energy Production Through Biomass Gasification Gas Technology: A Review, *F1000Research*, vol. 13, p. 511, doi: 10.12688/f1000research.147958.1
22 Elfaleh, I., Abbassi, F., Habibi, M., Ahmad, F. and Guedri, M. 2023. Review Article a Comprehensive Review of Natural Fibers and their Composites: An Eco-Friendly Alternative to Conventional Materials, *Results Eng.*, vol. 19, no. April, p. 101271, doi: 10.1016/j.rineng.2023.101271
23 Kumar, M., Tewari, R., Zafar, S. and Mavinkere, S. 2023. A Comprehensive Review of Various Factors for Application Feasibility of Natural Fiber-Reinforced Polymer Composites, *Results Mater.*, vol. 17, no. December, p. 100355, doi: 10.1016/j.rinma.2022.100355
24 Sujesh, G., Giri, A. D., Menon, A. and Nanditha, K. P. 2023. Hybrid Composites Manufacturing and Testing of Wind Turbine Blades Hybrid Composite Manufacturing, EasyChair preprints.
25 Sundar, D. 2017. Opportunities for Natural Fiber Reinforced Composites Towards Tropical Wind Turbine Material Needs, *2017 Asian Conference on Energy, Power and Transportation Electrification, ACEPT*, pp. 1–7.
26 Alabi, O. O., Adeaga, O. A., Ogunsiji, G. O. and Dada, S. A. 2023. Development and Performance Evaluation of Green-Vegetables Slicing Machine. *LAUTECH J. Eng. Technol.*, vol. 17, no. 2, pp. 40–46. https://laujet.com/index.php/laujet
27 Samuel, B. O., Sumaila, M. and Dan-Asabe, B. 2022. Physical and Mechanical Properties of Natural Fiber Reinforced Polymer Composites with Potentials for Wind Turbine Blade Applications: A Review, *Jurnal Mekanikal*, no. December, pp. 1–14.

28 Maurya, A., Bhattacharya, A. and Khare, S. K. 2020. Enzymatic Remediation of Polyethylene Terephthalate (PET). Based Polymers for Effective Management of Plastic Wastes: An Overview, *Front. Bioeng. Biotechnol.*, vol. 8, no. November, pp. 1–13, doi: 10.3389/fbioe.2020.602325
29 Satwika, N. A., Sarwono and Hantoro, R. 2018 Investigation Flow on Horizontal Axis Wind Turbine with Betz Chord Distribution, Twist, and Winglet, *Proceedings – 2018 4th International Conference on Science and Technology ICST 2018*, vol. 1, pp. 1–6, doi: 10.1109/ICSTC.2018.8528653
30 Hua, X., Zhang, C., Wei, J., Hu, X. and Wei, H. 2019. Wind Turbine Bionic Blade Design and Performance Analysis, *J. Vis. Commun. Image Represent.*, vol. 60, pp. 258–265, doi: 10.1016/j.jvcir.2019.01.037
31 Zhu, B., Sun, X., Wang, Y. and Huang, D. 2017. Performance Characteristics of a Horizontal Axis Turbine with Fusion Winglet, *Energy*, vol. 120, pp. 431–440, doi 10.1016/j.energy.2016.11.094
32 Mochane, M. J., Mokhena, T. C., Mokhothu, T. H., Mtibe, A., Sadiku, E. R. and Ray, S. S. 2019. Recent Progress on Natural Fiber Hybrid Composites for Advanced Applications: A Review, *Express Polym. J.*, vol. 13, no. 2, pp. 159–198.
33 Akil, H., Santulli, C., Sarasini, F., Tirillò, J. and Valente, T. 2014. Environmental Effects on the Mechanical Behavior of Pultruded Jute/Glass Fiber-Reinforced Polyester Hybrid Composites, *J. Compos. Sci. Technol.*, vol. 94, pp. 62–70.
34 Mohsin, M., Hanif, I. and Abbas, Q. 2021 Fueling the Future with Green Economy: An Integration of Its Determinants from Renewable Sources, *Energy*, p. 122029, doi: 10.1016/j.energy.2021.122029
35 Abbas, Q., Khan, A. R., Bashir, A., Alemzero, D. A., Sun, H. and Iram, R. 2020 Scaling Up Renewable Energy in Africa: Measuring Wind Energy Through Econometric Approach, *Environ. Sci. Pollut. Res.*, pp. 36282–36294.
36 Ahmed, M. M., Dhakal, H. N., Zhang, Z. Y., Barouni, A. and Zahari, R. 2021 Enhancement of Impact Toughness and Damage Behaviour of Natural Fiber Reinforced Composites and their Hybrids Through Novel Improvement Techniques: A Critical Review American Society of Testing Materials, *Compos. Struct.*, vol. 259, no. August, p. 113496, doi 10.1016/j.compstruct.2020.113496
37 Suriani, M. J., Rapi, H. Z. Ilyas, R. A. and Petr, M. 2021. Delamination and Manufacturing Defects in Natural Fiber-Reinforced Hybrid Composite: A Review, *Polymers*, pp. 1–24, doi: 10.3390/polym13081323
38 Manaia, P., Manaia, A. T., and Rodriges, L. 2019. Industrial Hemp Fibers: An Overview, *Fibers*, vol. 7, no. 12, pp. 1–16, doi: 10.3390/fib7120106
39 Gupta, M. K. and Srivastava, R. K. 2016. "Mechanical Properties of Hybrid Fibers-Reinforced Polymer Composite: A Review, *Polym. Plast. Technol. Eng.*, vol. 55, no. 6, pp. 626–642, doi: 10.1080/03602559.2015.1098694
40 Sinha, A. K., Narang, H. K. and Bhattacharya, S. 2017 Mechanical Properties of Natural Fiber Polymer Composites, vol. 37, no. 9, pp. 879–895, doi: 10.1515/polyeng-2016-0362
41 Salunkhe, S. 2022. In State of Art: Mechanical Behaviour of Natural Fiber Based Hybrid Polymeric In State of Art: Mechanical Behavior of Natural Fiber-Based Hybrid Polymeric Composites for Application of Automobile Components, no. March, doi: 10.1002/pc.26045
42 De Queiroz, H. F. M., Aguiar, R. A. A. and Banea, M. D. 2021. A Review on the Thermal Characterisation of Natural and Hybrid Fiber Composites, *Polymers*, doi: 10.3390/polym13244425

43 Ibrahim, T. K., Yawas, D. S., Dan, B. and Adebisi, A. A. 2023. Taguchi Optimization and Modeling of Stir Casting Process Parameters on the Percentage Elongation of Aluminum, Pumice, and Carbonated Coal Composite, *Sci. Rep.*, pp. 1–14, doi: 10.1038/s41598-023-29839-8
44 Akoredeley, A., Christian, L., Adeniyi, K., Okon, B. and Uhuotu, M. 2022. Powder Production from Recycled Low-Density Polyethylene Waste and Doum Palm Nuts for Light Weight Engineering Application: A Circular Economy Approach, *Clean. Waste Syst.*, vol. 3, no. October, p. 100044, doi: 10.1016/j.clwas.2022.100044
45 Alabi, O. O., Adeaga, O. A. and Dare, A. A. 2024. Reviewing the Enhancement Optimum Performance Characteristics of Horizontal Axis Wind Turbine Blades Using Add-on of Suitable Aerodynamic Properties, vol. 5, no. 1, pp. 93–99, doi: 10.14744/thermal.0000848
46 Rahimizadeh, A., Kalman, J., Fayazbakhsh, K. and Lessard, L. 2019. Recycling of Fiberglass Wind Turbine Blades into Reinforced Filaments for Use in Additive Manufacturing, *Compos. Part B*, vol. 175, no. February, p. 107101, doi: 10.1016/j.compositesb.2019.107101
47 Rani, M., Choudhary, P., Krishnan, V. and Zafar, S. 2021. A Review on Recycling and Reuse Methods for Carbon Fiber/Glass Fiber Composites Waste from Wind Turbine Blades, *Compos. Part B*, vol. 215, no. October p. 108768, doi: 10.1016/j.compositesb.2021.108768
48 Cousins, D., Suzuki, S., Murray, Y. and Samaniuk, J. R. 2019. Recycling Glass Fiber Thermoplastic Composites from Wind Turbine Blades, *Compos. Part B*, vol. 175, pp. 1–28, doi: 10.1016/j.compositesb.2019.107101
49 Musial, W. 2018. Trends in the Design, Manufacture and Evaluation of Wind Turbine Blades, no. November, doi: 10.1002/we.90
50 Javaid, U., Khan, Z. M., Khan, M. B., Bassyouni, M. and Hasan, S. W. 2016. Fabrication and Thermo-Mechanical Characterization of Glass Fiber/Vinyl Ester Wind Turbine Rotor Blade, *Compos. Part B*, vol. 2016, doi: 10.1016/j.compositesb.2015.12.034
51 Grujicic M. 2010. Multidisciplinary Design Optimization for Glass-Fiber Epoxy-Matrix Composite 5 MW Horizontal-Axis Wind-Turbine Blades, vol. 19, no. November, pp. 1116–1127, doi: 10.1007/s11665-010-9596-2
52 Srivastava, A. 2018. Preparation and Mechanical Properties of Jute Fiber Reinforced Epoxy Industrial Engineering & Management Preparation and Mechanical Properties of Jute Fiber Reinforced Epoxy Composites, no. November, doi: 10.4172/2169-0316.1000234
53 Vidyashri, V. 2019 Preparation of Chemically Treated Sugarcane Bagasse Fiber Reinforced Epoxy Composites and their Characterization Preparation of Chemically Treated Sugarcane Bagasse Fiber Reinforced Epoxy Composites and their Characterization, *Cogent Eng.*, doi: 10.1080/23311916.2019.1708644
54 Lomwongsopon, P. and Varrone, C. 2022. Critical Review on the Progress of Plastic Bioupcycling Technology as a Potential Solution for Sustainable Plastic Waste Management, *Polymers (Basel)*, vol. 14, no. 22, doi: 10.3390/polym14224996
55 Sharma, H. 2023. Critical Review on Advancements on the Fiber- Reinforced Composites: Role of Fiber/Matrix Modification on the Performance of the Fibrous Composites, *J. Mater. Res. Technol.*, vol. 26, pp. 2975–3002, doi: 10.1016/j.jmrt.2023.08.036

56 Renard, E., Rademacher, T., Hu, F., Demleitner, M. and Mattar, N. 2020. Fully Bio-Based Epoxy-Amine Thermosets Reinforced with Recycled Carbon Fibers as a Low Carbon-Footprint Composite Alternative, no. January, doi: 10.1021/acsapm.0c01187

57 Abir, N., Bakr, A., Ara, H., Gafur, A., Nabi, A. and Mahmud, A. 2023 Heliyon Effect of Fibre Loading on Mechanical Properties of Jute Fibre Bundle Reinforced Gypsum Composites, *Heliyon*, vol. 9, no. 7, p. e18147, doi: 10.1016/j.heliyon.2023.e18147

58 Balogun, O. A., Akinwande, A. A., Adesina, O. S. and. Olasoju, O. S. 2020. Heliyon in Fluence of Chemical Treatment on the Properties of Cement-Paper Hybrid Composites for Ceiling Board Application, *Heliyon*, vol. 6, no. June, p. e04512, doi: 10.1016/j.heliyon.2020.e04512

59 Krzyzak, A. and Kosicka, E. 2020. Research into the Effect of Grain and the Content of Alundum on Tribological Properties and Selected, *Materials*, vol. 13, no. 24, p. 5735, doi: 10.3390/ma13245735

60 Ali, Y., Sabeeh, N. N., Yass, M. F. and Salih, E. 2019. Concrete Beams Strengthened with Jute Fibers Concrete Beams Strengthened with Jute Fibers, no. April, doi: 10.28991/cej-2019-03091286

61 Saraswati, P. K., Sahoo, S., Parida, S. P. and Jena, P. C. 2019. Fabrication, Characterization and Drilling Operation of Natural Fiber Reinforced Hybrid Composite With Filler (Fly-Ash/Graphene), no. August, doi: 10.35940/ijitee.J1253.0881019

62 Shelar, P. B. 2021. A Short Review on Jute Fiber Reinforced Composites A Short Review on Jute Fiber Reinforced Composites, no. July, doi: 10.4028/www.scientific.net/MSF.1019.32

63 Comparisons, T. 2022. Selection of Automotive Brake Material Using Different MCDM Techniques and Their Comparisons, vol. 15, no. 1, pp. 24–33, doi: 10.25103/jestr.151.04

64 Czajka, A., Bulski, R., Iuliano, A., Plichta, A., Mizera, K. and Ryszkowska, J. 2022. Grafted Lactic Acid Oligomers on Lignocellulosic Filler towards Biocomposites, *Materials*, vol. 15, no. 1, doi: 10.3390/ma15010314

Chapter 5

Green composites

Advancing materials, processing, and applications with emphasis on machine learning (ML) and artificial neural networks (ANN) integration

Prasanjit K. Dey and Arup R. Bhattacharyya

5.1 INTRODUCTION

Advancement of sustainable development is now emerging and reducing the impact on the environment since the significant development in the research of green composites become a milestone in materials science and engineering. The combination of two or more different materials results in the composites yielding a more superior property. They are known for their increased strength-to-weight ratio, durability, toughness, stiffness, and resistance to corrosion, impact, wear, and fire. The presence of composites dates back to 3400 B.C, when the plywoods were created by gluing the wood strips at different angles by the Mesopotamians. With the scientific advancements, significant growth has taken place in the field of synthetic materials development for composites. In the year 1930, for the very first time, glass fibers and polymer resins were introduced. Subsequently, in 1960s, the emergence of carbon fibers paved the way for the development of modern fiber-reinforced polymer (FRP) composites, as shown in Figure 5.1. These green composites have emerged as a sustainable solution for the growing economy (Maiti et al., 2022). Green composites composed of natural fibers and biodegradable resins offer a significant reduction in the environmental impact compared to their traditional counterparts.

Most of the traditional composites are made of synthetic fibers and petroleum-based resins, which contributes to environmental pollution and resource depletion. However, green composites are made of renewable resources to foster sustainable development. Natural fibers such as flax, hemp, jute, kenaf, sisal, coir, etc. are used to formulate green composites with biodegradable matrices such as polylactic acid (PLA), polyhydroxyalkanoates (PHAs), starch-based polymers, soy protein-based resins, etc. The selection of natural fibers and matrix is dependent on their applications, and these composites are mechanically comparable with the synthetic composites and also has environmental advantage. Natural fibers are cheap, biodegradable, and have excellent specific strength and stiffness, making them excellent alternatives to synthetic fibers. On the other hand, the use of biodegradable matrices has lots of advantages such that composites can

DOI: 10.1201/9781003527275-5

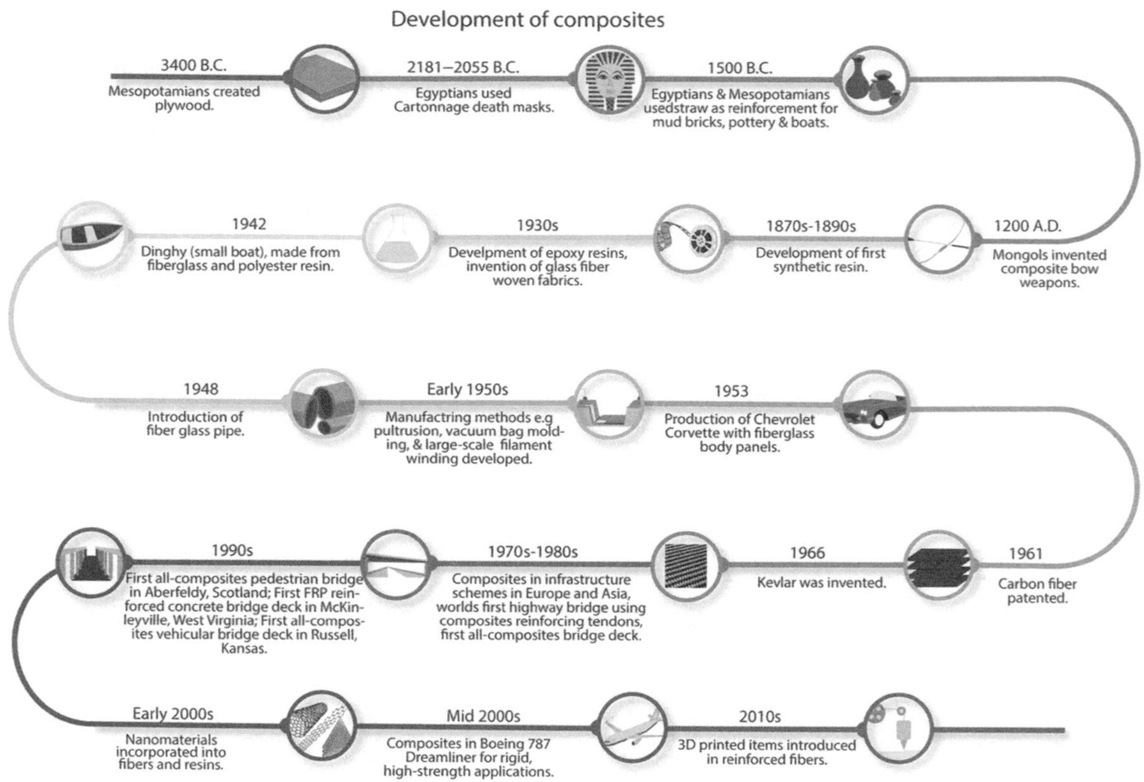

Figure 5.1 Evolution of composites manufacturing for several industries.

decompose naturally, thereby reducing waste and environmental pollution and fulfilling the sustainable development goals (SDG 11 and 12).

One of the primary motivations is the rising concern over environmental degradation and pollution. The production and disposal of traditional composite materials contribute significantly to pollution, greenhouse gas emissions, and landfill waste. Green composites address these issues by being derived from renewable resources and being biodegradable, which means they have a much lower environmental footprint.

Traditional composites also lead to the depletion of fossil resources, and this leads to unsustainable practices in the long term due to the finite nature of fossil resources. Plant-based and biodegradable materials are renewable in nature, and being more sustainable become the best alternative to traditional composites. This promotes the use of agricultural by-products and waste, reduces the consumption of fossil resources, and contributes to a circular economy (Rowell & Rowell, 1996). The use of green composites also mitigates the adverse environmental impacts associated with traditional composites. As industries and governments worldwide adopt more stringent environmental regulations and sustainability goals, the demand for green composites is expected to rise.

The aim of this chapter is to give a comprehensive overview of the development of green composites and explore the natural fibers and biodegradable polymer matrices used in these composites, and this chapter also discuss on the processing techniques employed in the manufacturing of green composites, such as injection molding, compression molding, resin transfer molding (RTM), extrusion, and additive manufacturing – highlighting their advantages and applications. Understanding of various techniques and processes is crucial for optimizing the performance and sustainability of green composites (Mulenga et al., 2021). Diverse applications of green composites across various industries, including automotive, construction, packaging, consumer goods, aerospace, and biomedical fields, are also discussed. In addition to traditional applications, the chapter will discuss the integration of cutting-edge technologies such as machine learning (ML) and artificial neural networks (ANN). These technologies are revolutionizing the development and optimization of green composites by enabling better material design, process optimization, quality control, and performance prediction. The use of ML and ANN in green composites represents a significant advancement in the field, offering new possibilities for innovation and efficiency.

5.2 MATERIALS AND COMPOSITION

The foundation of green composites lies in their materials – biodegradable and renewable resources that offer a sustainable alternative to conventional composites. This section will discuss the various natural fibers and biopolymers used in green composites, their properties, and recent advancements in material development. Also, various surface treatments for natural fibers are discussed to enhance the fiber-matrix adhesion.

5.2.1 Natural fibers

Natural fibers act as a reinforcement for green composites and thus improve the mechanical properties of these materials. Most commonly used natural fibers are flax, hemp, jute, sisal, and bamboo. Each type of fiber has distinct characteristics, making them suitable for a variety of applications.

Flax: Flax fibers are extracted from the stem of the flax plant (*Linum usitatissimum*). Flax fibers are known for their high tensile strength and stiffness, and they are widely used in automotive and construction applications. They have low density, which makes them suitable for lightweight composite structures. This fiber has excellent mechanical properties, such as high specific modulus and specific strength, making it a preferred choice for reinforcing composites. The processing of this fiber is relatively easy and can be woven into fabrics for enhanced structural applications. H. Li et al. (2022) have discussed comprehensively on the preparation of flex reinforced composites and their multifunctional applications. A study by Bajpai et al. (2021) demonstrated that flax-reinforced composites exhibit superior impact resistance and dimensional stability, making them ideal for structural applications.

Hemp: Hemp fibers are derived from the *Cannabis sativa* plant. Their mechanical properties are excellent and are highly sustainable. They have good tensile strength, durability, and resistance to degradation, hence making them suitable for applications that demand high strength and durability. Hemp cultivation is environmentally friendly and requires minimal pesticides and herbicides. They have a wide range of applications, including automotive parts, textiles, and bioplastics. Their natural resistance to mold and UV light further enhances their utility in outdoor applications. The author Zimniewska (2022) extensively discussed the processing and properties of hemp fibers. He highlighted the prospects of the hemp fibers as one of the important raw materials of textiles. Sheedev Antony et al. (2021) investigated the fiber volume fraction effect of hemp fiber woven fabrics/polypropylene (PP) composite laminates on the tensile properties and impact strength. Research by Tutek and Masek (2022) highlighted the potential of hemp fibers in biodegradable composites for packaging and consumer products, emphasizing their environmental benefits.

Jute: Jute fibers are obtained from the bark of the jute plant (Corchorus species). They are cheap and easily available, making them suitable for low-cost applications. They are most commonly used in packaging, textiles, and automotive industries. Their tensile strength and stiffness are moderate but offer excellent sound and thermal insulation properties. Jute is biodegradable and compostable, contributing to its appeal as a sustainable material. The cultivation of jute is beneficial for the soil as it enhances soil fertility. Vikky Rajulapati et al. (2020) observed higher tensile strength and Young's modulus for jute fiber treated with an enzyme mixture than untreated and were similar to those of NaOH treated. They showed that the *Ct*PME and *Ct*PL1B mixture can be used as a viable green bioprocess instead of a chemical process in the textile industry.

Sisal: The leaves of *Agave sisalana* are the source of sisal fibers. They are known for their good mechanical properties and are most commonly used in construction and automotive sectors. It has high tensile strength and is durable under various environmental conditions, due to excellent wear and tear resistance. Sisal fibers are also resistant to microbial attack and have good frictional properties. These characteristics make sisal suitable for use in ropes, mats, and reinforcement in polymer composites. The mechanical properties of sisal FRP composites have been thoroughly reviewed and studied by Ibrahim et al. (2015).

Kenaf: Kenaf fibers are derived from the plant *Hibiscus cannabinus* and are known for their high specific strength, making them valuable in automotive and construction industries. The kenaf fibers are also used in paper production and as soil conditioners. They are lightweight, have higher tensile strength, and are biodegradable. They offer excellent thermal insulation and acoustic properties, making them ideal for building materials. It is one of the fast-growing plants that needs minimal agricultural inputs, thus making it one of the sustainable fiber sources. Studies have demonstrated the effectiveness of kenaf fibers in various applications (Akil et al., 2011). Datta and Kopczyńska (2015) examined the morphology and mechanical properties of biocomposites obtained from kenaf natural fiber and thermoplastic polyurethane. The kenaf fiber finds lots of applications, such as in paper production, as a soil conditioner, etc.

Coir: The fibers extracted from the husk of coconuts (*Cocos nucifera*) are known as coir fibers. They are highly durable and versatile which makes them suitable for a variety of applications. Their strength and natural resistance to saltwater and microbial deterioration make them suitable for usage in packaging and automotive industries. Coir can be helpful in erosion control and soil stabilization. Geotextiles and agricultural products benefit from the fibers' toughness and resilience, as well as their high elongation at break. Ali et al. (2012) found that coir fibers can improve the mechanical properties of composite materials used in automobile components, highlighting their potential for increased durability and weight reduction. Also, Dash et al. (2015) investigated the use of coir in geotextiles and discovered that coir's

Table 5.1 Mechanical properties of different natural fibers compared to conventional reinforcing fibers

Fiber		*Density [g cm^{-3}]*	*Tensile strength [MPa]*	*Young's modulus [GPa]*	*Elongation at break [%]*	*Equilibrium moisture content [%]*
Eco-friendly sustainable natural fibers	Flax	1.54	345–2000	27–85	1–4	7
	Hemp	1.47	368–800	17–70	1.6	9
	Jute	1.44	393–773	10–30	1.5–1.8	12
	Sisal	1.33	600–700	38	2–3	11
	Kenaf	1.2	240–930	14–53	1.6	—
	Coir	1.25	220	6	15–25	10
	Bamboo	0.6–1.1	140–230	11–17	—	8.9
Conventional fibers	E-glass	2.5	2000–3500	70–77	4.5–4.9	—
	S-glass	2.5	4570	86–90	4.5–4.9	—
	Carbon	1.8–1.9	3400–5400	230–440	1.4–1.8	—
	Aramid	1.45	3400–4000	130–185	2.5	—

strong tensile strength and biodegradability considerably increase soil stabilization and erosion management. In addition, Mishra et al. (2008) conducted a study that assessed the use of coir fibers in agricultural goods. The study has shown that coir fibers have the capacity to improve soil aeration and water retention, hence facilitating more robust plant growth. Comparison of the mechanical properties of various natural fibers with conventional reinforcing fibers is shown in Table 5.1.

Bamboo: Bamboo fibers are becoming more and more popular because of their quick growth and sustainability. They are appropriate for structural applications due to their high strength and stiffness. Bamboo fibers have the potential to replace conventional building materials, according to a study by Zhang et al. (2022) that examined their application in green composites for building materials. The study showed that adding bamboo fibers to composite materials improves their mechanical qualities while also lowering the environmental impact of building materials. As a result, bamboo fibers present a viable substitute for environmentally friendly building techniques.

5.2.2 Biodegradable matrices

Biopolymers are one of the popular classes of functional materials that are well suited for high-value uses. Several efforts have been made to substitute synthetic polymers with biodegradable materials, particularly those produced from natural resources, in order to lower the impact on the

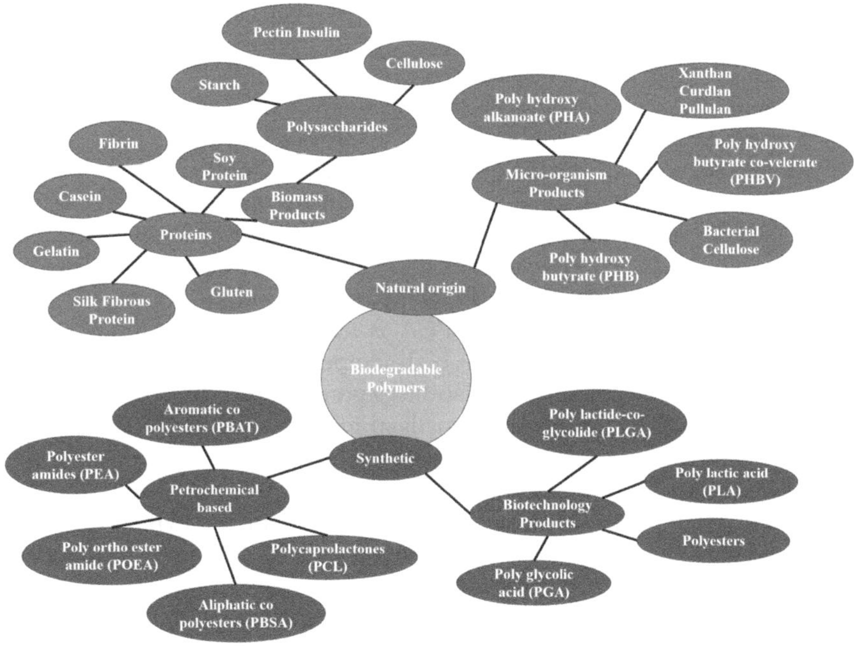

Figure 5.2 Classification of biodegradable polymers.

environment and the reliance on fossil fuels. These polymers are derived from renewable sources and are biodegradable, hence enhancing the sustainability of the composite material. Some of the common biopolymers are polylactic acid (PLA), polyhydroxyalkanoates (PHA), polyglycolic acid (PGA), and polybutylene succinate (PBS). The classification of biodegradable polymers is given in Figure 5.2.

Polylactic acid (PLA): One of the most widely used biopolymers in green composites is polylactic acid (PLA). It comes from sustainable sources like corn and sugarcane and is renowned for having good mechanical qualities and being biodegradable. The usage of PLA in green composites for packaging applications was investigated by X. Li et al. (2023) and their research showed that PLA has the potential to replace traditional plastics. Since PLA is a thermoplastic aliphatic polyester, it can be processed using conventional plastic processing techniques, and thus it makes PLA a versatile and excellent choice for the manufacturing of green composites.

Polyhydroxyalkanoates (PHA): PHA is produced by microbial fermentation and is a family of biopolymers. They are excellent in biodegradability and biocompatibility, making them ideal for medical and packaging applications. Chen et al. (2023) studied the usage of PHA in green composites for biomedical applications, demonstrating their promise in tissue engineering and drug delivery systems. PHAs are polyesters that accumulate in microbial cells as energy storage molecules. They may be synthesized from a variety of carbon sources, including waste materials, making them an environmentally benign choice as a biodegradable matrix.

Polyglycolic acid (PGA): Polyglycolic acid is a biodegradable polymer with high strength and excellent thermal stability. It is widely used in medical applications including sutures and implants. Joshi et al. (2021) investigated the application of PGA in green composites for medical devices, emphasizing their biocompatibility and biodegradable properties. The melting point of PGA is high and is a highly crystalline polymer. PGA can be used in high-temperature applications due to its higher melting point. It is also biodegradable and biocompatible and can be used as a temporary medical implant that gradually degrades within the body.

Polybutylene succinate (PBS): PBS is derived from renewable sources and is a biodegradable polymer with excellent thermal stability and mechanical properties. It is commonly used in packaging and agricultural applications. PBS is a thermoplastic polyester, and hence it can be processed by conventional plastic processing methods. Kim et al. (2021) examined the usage of PBS in green composites for packaging materials, revealing its potential to replace traditional plastics. Its biodegradability and versatility make it suitable for a wide range of applications, including films, fibers, and molded products.

Starch-based polymers: Starch-based polymers, produced from natural starch from plants such as corn, potatoes, and tapioca, are totally biodegradable, dissolve in water and carbon dioxide under appropriate environmental circumstances. Their mechanical characteristics can be modified by mixing them with different polymers or plasticizers; they typically offer good flexibility and tensile strength, while enhancing water resistance frequently requires modification. The lower melting point of these polymers and susceptibility to moisture may affect the processing and thermal stability. They can be processed using conventional plastic processing such as blowing films, extrusion, and injection molding. In a study, Avérous and Pollet (2012) found that starch-based polymers are a popular substitute for conventional plastics in packaging applications because they can be composted, making them suitable for usage in biodegradable bags, films, and food containers.

Soy protein-based resins: Soybeans are the main source of soy protein-based resins. These are biodegradable, sustainable, and eco-friendly alternatives to petroleum-based resins. They have very good mechanical properties such as tensile strength and elasticity. However, they may need modifications to improve water resistance and durability. They are moderate in thermal stability, and this can be improved using additives in high-temperature applications. These resins can be processed using traditional plastic manufacturing methods, such as injection molding and extrusion. In a study by Kurose et al. (2014), soy protein-based resins are used in packaging materials, including films, trays, and containers, where the capacity to degrade naturally is a major benefit. These resins are also explored for use in interior panels, trim components in automotive industry, offering an eco-friendly alternative to traditional plastics, as highlighted by Mungroo et al. (2014). Also, soy protein-based resins are utilized in adhesives and coatings, providing a sustainable option for various industrial applications.

Chitosan: Chitosan is a natural polymer obtained from chitin, found in the exoskeletons of crustaceans such as shrimp, crabs, and lobsters. The polymer chitosan is obtained after the deacetylation of chitin, making it one of the most available and abundant biopolymers. This biopolymer is of interest due to its unique properties and wide range of applications across various industries. The key properties of chitosan are biodegradability, biocompatibility, antimicrobial activity, film-forming ability, and solubility in acidic solutions. These properties make chitosan a sustainable alternative to synthetic polymers and suitable for various uses. In the biomedical field, chitosan is used for wound dressings, tissue engineering, and drug delivery systems as it has the ability to promote wound healing and it is biocompatible in nature (Jayakumar et al., 2010). It can serve as a drug carrier in controlled-release formulations in pharmaceutical applications, enhancing drug bioavailability and protecting against degradation (Felt et al., 1998). In addition, chitosan is also used as a food preservative in food industry, leveraging its antimicrobial properties to extend the shelf life of perishable goods (No et al., 2007). Also, this polymer is effective in water purification, binding with pollutants to remove heavy metals, dyes, and other contaminants from water (Rinaudo, 2006). In agriculture, chitosan acts as a natural pesticide and plant growth enhancer, promoting seed germination, enhancing disease resistance, and improving crop yield (El Hadrami et al., 2010). This chitosan polymer has diverse applications due to its versatility and eco-friendly across multiple sectors.

5.2.3 Treatments to enhance fiber-matrix compatibility

Natural fibers form a poor interface with polymer matrices and hence often require treatment to improve their compatibility with polymer matrices. These treatments lead to improvement in the adhesion between fibers and the matrix, which forms a strong interface resulting in increased fiber strength and lower moisture absorption. Andrew & Dhakal (2022) discussed several methods that are commonly used to treat the natural fibers, and such various methods are shown in Figure 5.3.

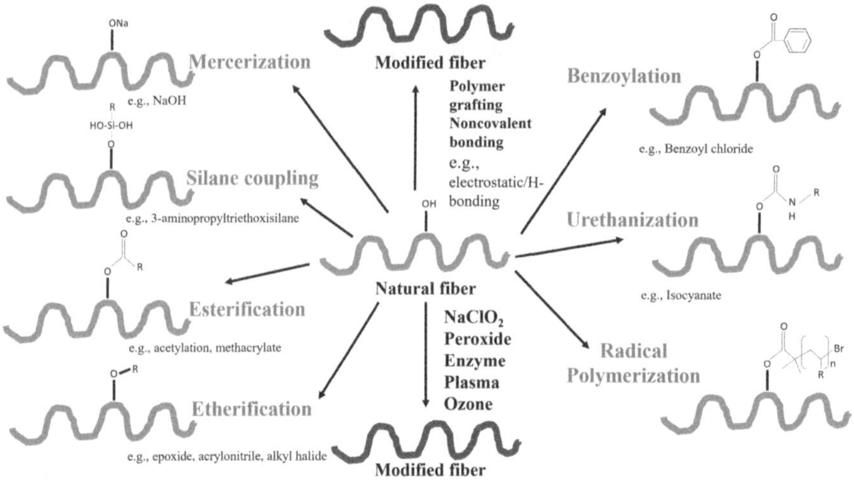

Figure 5.3 Illustration of various methods of surface treatments and modifying biofibers.

Alkali treatment: Alkali treatment, also known as mercerization, involves immersing the fibers in an NaOH solution for a specific duration, followed by thorough washing and drying. This alkali treatment removes hemicellulose, lignin, and other impurities from the fiber surface and results in enhancement of surface roughness and improving fiber-matrix adhesion. The crystallinity of the cellulose in the fibers is increased after alkali treatment, which improves the mechanical properties.

Silane treatment: Silane treatment is the process of improving the interfacial interaction between natural fibers and polymer matrices by using silane coupling agents. The functional groups of the silane molecule can react with both the fiber surface and polymer matrix, forming strong covalent bonds. This strong adhesion between the fibers and the matrix results in improved mechanical properties of the composite. This treatment also results in lower water absorption, due to which the durability of the composites enhances in humid environments.

Acetylation: This is a chemical modification process in which acetic anhydride is treated with natural fibers known as acetylation. This treatment substitutes the hydroxyl groups in the cellulose with acetyl groups, reducing the hydrophilicity of the fibers. As a result, the absorption of moisture by acetylated fibers is reduced, and this improves their dimensional stability and resistance to microbial attack. This process of acetylation also improves the fibers' compatibility with hydrophobic polymer matrices, which improves the composite's adhesion and dispersion.

Other treatments: There are various other methods used for the treatment of natural fibers apart from those discussed above, including enzyme treatment, plasma treatment, and graft copolymerization, which are also employed to modify natural fibers. In an enzyme treatment, specific enzymes are used to selectively remove non-cellulosic components from the fibers, resulting in enhancement of their mechanical properties. Plasma treatment modifies the fiber surface at a molecular level, introducing functional groups that improve fiber-matrix bonding. Grafting polymer chains onto the fiber surface involves the graft copolymerization process, increasing compatibility with the polymer matrix. These various treatments played an important role in the improvement of the bonding strength between natural fibers and biopolymer matrices, resulting in the optimization of the performance of green composites. This enhancement of fiber-matrix adhesion, increased fiber strength, and reduced moisture absorption of treated natural fibers contribute to the development of composites, which are sustainable, eco-friendly, and meet the performance criteria. Depending on the type of natural fiber, the desired qualities of the composite, and the particular application requirements, the right treatment procedure selection is made. Recent studies on surface-treated lignocellulosic fiber-reinforced biocomposites are given in Table 5.2.

Table 5.2 Recent studies on surface-treated lignocellulosic fiber-reinforced biocomposites

Treatment methods	*Composites*	*Results*
Chemical treatments		
NaOH	Hemp/euphorbia	Increase in shear strength, tensile strength
	Sisal/polyester	Increase in tensile strength
Silane & NaOH	Flax/PLA	Improved hydrophobicity and mechanical properties
	Hemp/PLA	Adhesion force enhanced
Mercerization	Sisal/PP	Increase in tensile, flexural, and impact strengths
	Jute/epoxy	Transverse composite strength improved
	Flax/epoxy	Tensile strength and modulus improved
Physical treatments		
Argon and air-plasma	Wood/PP	Increased tensile strength and modulus
Autoclave (0.5–2 bar)	Flax-organic resin	Enhanced moisture resistance
Laser, ozone, and plasma	Jute	Increased surface roughness
UV radiation (up to 50)	Jute/PE/PP	Enhanced tensile and flexural properties

(*Continued*)

Table 5.2 (Continued)

Treatment methods	*Composites*	*Results*
Biological treatments		
Enzyme	Flax/hemp	Improved thermal stability
Schizophyllum commune (fungi)	Hemp/PP	Improved tensile strength (28%)
Fungal treatment	Hemp/PP	22% higher strength

5.3 PROCESSING TECHNIQUES

The processing of green composites plays a crucial role in determining their properties and performance (Maiti et al., 2022). This section will discuss various processing methods, recent advancements, and the impact of processing parameters on the final properties of green composites. Schematics of common composite processing techniques are shown in Figure 5.4.

5.3.1 Hand lay-up

Hand lay-up is a simple and cost-effective way of making fiber-reinforced composites. Layers of reinforcing fibers are manually put into a mold with brushes or rollers, followed by the deposit of a resin matrix. After the glue has been wetted through the fibers, the laminate is either allowed to cure at room temperature or heated. Hand lay-up is a good technique for specialized and low-volume applications since it can construct flax fiber-reinforced epoxy composites well, per a study by Barouni et al. (2022). The mold must be prepared before beginning the process, and the fiber reinforcements must be cut to appropriate size and shape. Next, these fibers are layered and added to the mold, followed by an equal coating of resin. Until the desired thickness is achieved, lay-up is done again. The laminate is allowed to solidify once all of the layers are in place, producing a sturdy and rigid composite structure. Hand lay-up is used because it can be used to manufacture a large variety of shapes and sizes and has a low initial cost. It is ideal for small-scale production runs and prototyping because of its quick modification and tweaking capabilities. Furthermore, because it doesn't require complicated machinery, smaller companies and workshops may complete the procedure. Hand lay-up is labor- and time-intensive, though, which lowers its efficiency in large-scale production. The end product's consistency and quality can also be impacted by the operator's skill level. Despite these disadvantages, hand lay-up is a widely used method for producing custom, high-grade composite parts.

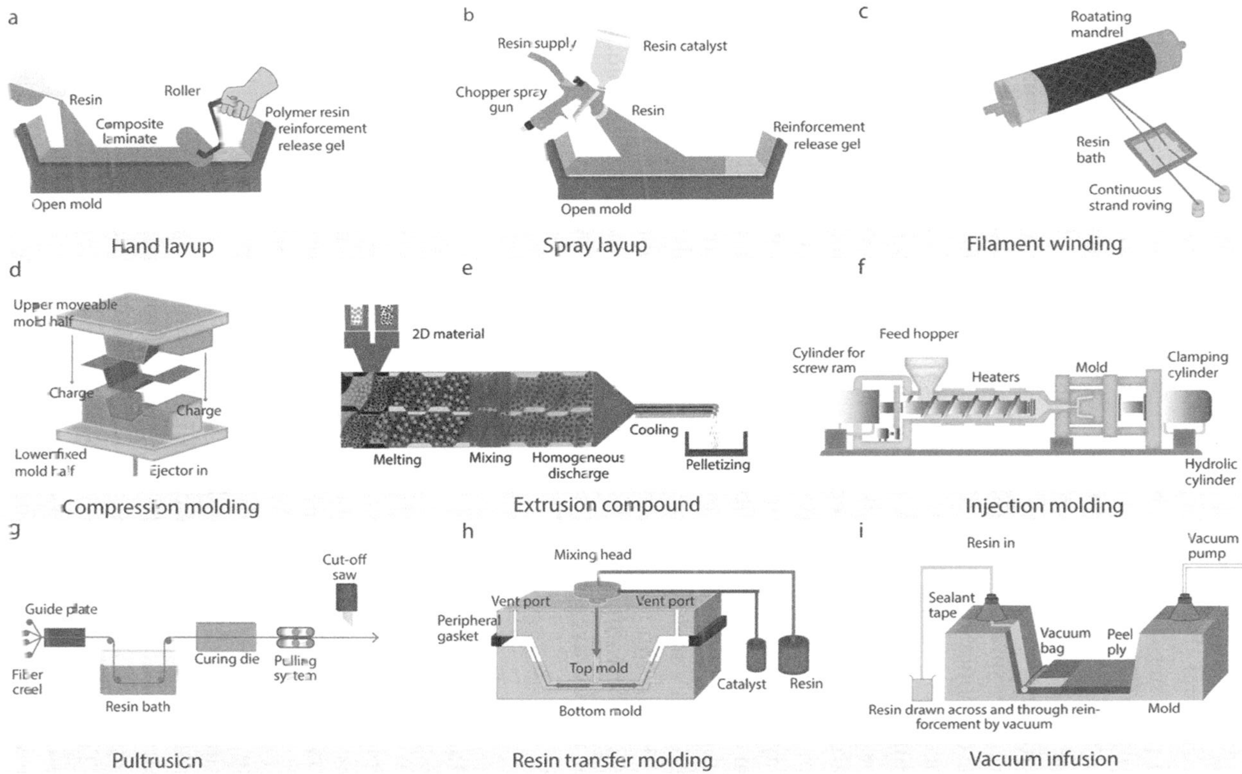

Figure 5.4 Diagrams of typical composite fabrication methods, (a) hand lay-up, (b) spray lay-up, (c) filament winding, (d) compression molding, (e) extrusion compound, (f) injection molding, (g) pultrusion, (h) RTM, and (i) vacuum infusion or vacuum assisted resin transfer.

5.3.2 Spray lay-up

Similar to hand lay-up, spray lay-up is a manufacturing procedure that entails spraying a resin and chopped fiber mixture onto a mold surface. Using this method, enormous, lightweight composite structures with detailed forms and shapes can be created. In order to fabricate glass fiber-reinforced polyester composites, Singh et al. (2020) investigated the use of spray lay-up and showed that it was a useful technique for creating big panels and shells. Making the mold and choosing the fiber and resin components are the first steps in the process. Short lengths of fiber are cut, and then they are combined with resin in a spray gun. Even layers of this mixture are subsequently sprayed onto the mold surface. To ensure adequate consolidation, the laminate can be compacted and air bubbles removed using rollers. The laminate is allowed to cure and form a stiff composite structure after reaching the required thickness. Spray lay-up has many benefits, one of which is its capacity to quickly and labor-efficiently create big, intricate shapes. Large-scale, inexpensive components like boat hulls, car panels, and architectural structures can be produced with this technology. Besides, the use of spray lay-up offers versatility in material selection by permitting the integration of different kinds of fibers and resins. Variations in thickness and fiber distribution, however, may arise from the procedure and could impact the final product's structural performance. There are also health and environmental issues with the usage of volatile organic compounds (VOCs) in resins. Despite these difficulties, spray lay-up is still a widely used technique because of its effectiveness and adaptability when creating big composite structures.

5.3.3 Filament winding

Filament winding is used to create high-strength, hollow composite structures like pipes, tanks, and pressure vessels. This process involves wrapping continuous resin-impregnated fibers around a rotating mandrel in specific patterns. The process begins with the selection of continuous fiber tows, which are passed through a resin bath where they are impregnated with resin. These resin-saturated fibers are then wound around a rotating mandrel in a predetermined pattern, often involving helical or hoop winding angles. The wound structure is subsequently cured, either at room temperature or with the application of heat, to solidify the resin and form a rigid composite part. After curing, the mandrel is removed, leaving a hollow, seamless composite structure. The application of filament winding to manufacture carbon fiber-reinforced epoxy composite pressure vessels was investigated by (Martin, 2024), showcasing the technique's ability to produce lightweight, high-performance components. This method provides numerous benefits, such as the production of high-strength, lightweight components with exceptional structural integrity and uniformity. Automation in process enables precise control over fiber placement and resin content, confirming consistent properties throughout the components. Filament-wound composites are extensively utilized in aerospace, automotive, and industrial sectors, where high strength-to-weight ratios and resistance to internal pressures are essential. However, this technique is very limited to producing cylindrical or spherical shapes and generally cannot accommodate complex geometries.

5.3.4 Compression molding

Making green composites is a common application for compression molding. To create the composite material, the fiber-matrix combination is put into a mold, then heated and compressed. With this technique, fiber orientation and distribution may be precisely controlled, producing composites with consistent qualities. Compression molding was used to create flax-reinforced PLA composites, as shown in the research by Patel et al. (2023), highlighting the materials' exceptional mechanical qualities and biodegradability. The fiber-matrix mixture is made first and then poured into a heated mold to start the process. Following the closure of the mold, pressure is used to mold the material into the required shape. When heat and pressure are applied, the biopolymer melts and flows, encasing the natural fibers and solidifying into a composite when it cools. Large, flat, and generally simple components like panels, sheets, and structural pieces are especially well suited for manufacture via compression molding. Producing components with a high fiber content and superior mechanical characteristics is just one of the process's many benefits. Also, high fiber alignment composites can be produced using compression molding, increasing the material's strength and stiffness. Precise control over processing parameters, including temperature, pressure, and molding time, is a primary difficulty in compression molding.

5.3.5 Extrusion

Extrusion is a continuous manufacturing technique that creates uniformly long profiles. To create the composite material, the process involves pushing the fiber-matrix mixture through a die. Green composites in the form of sheets, films, and pipes can be produced using this technique. Mochane et al. (2021) investigated the use of extrusion to produce bamboo-reinforced PBS composites, emphasizing their potential for construction and packaging applications. The process starts with preparing the fiber-matrix mixture, which is then put through an extruder. The mixture is heated and pressed through a die to create a continuous profile, which is then cooled and cut to the correct length. Extrusion has various advantages, including the capacity to create lengthy and continuous profiles of consistent quality. The technique is highly automated, and it can generate components with complex cross-sections. Extrusion also enables the creation of composites with high fiber content and superior mechanical characteristics. Extrusion presents a number of difficulties, the leading among them being the requirement for exact control over processing variables like temperature, pressure, and cooling rate. The quality and functionality of the finished product can be greatly impacted by these factors. The development of specialized dies and equipment for the process can be costly and time-consuming.

5.3.6 Injection molding

High-speed production techniques like injection molding are utilized to create elaborate designs and complicated shapes. High pressure is used to inject the fiber-matrix combination into a mold cavity. This process delivers superior dimensional accuracy and is appropriate for producing green composites in large quantities. The potential for high-volume applications of hemp-reinforced PHA composites was demonstrated by Dahal et al. (2022) through their investigation into the utilization of injection molding in this process. The biopolymer is heated to a fluid state to start the process. After that, the melted polymer is combined with natural fibers to create a uniform combination. After being injected into a mold cavity, the liquid solidifies and cools, taking on the shape of the mold. High precision is possible when producing intricate and delicate components using injection molding. The capacity of injection molding to swiftly and reliably produce vast volumes of parts of consistent quality is one of its key benefits. The procedure can create components with fine features and close tolerances and is highly automated. Injection molding is a flexible technique and applies to many natural fibers and biopolymers. Thus, green composites can be tailored to fit the needs of certain applications.

5.3.7 Pultrusion

To produce long, straight composite profiles with homogeneous cross-sections, pultrusion is a continuous production technique. Through a resin bath and a heated die, where the resin dries and hardens, reinforcing fibers are drawn through this technique. Carbon FRP (CFRP) rods were made via pultrusion, and the work by Guo et al. (2022) examined this process and shown how it can yield lightweight, highly strength structural elements. Prior to being impregnated with resin, a resin bath is used to feed a selection of reinforcing fibers from creels. Once through a heated die, the fibers drenched in resin are formed and allowed to cure to create the composite profile. Cut to the desired length, the cured profile is constantly drawn out. Not only does pultrusion generate lightweight, high-strength components with great dimensional stability and consistency, but it also has other advantages. Because of the high level of automation in the process, less labor is needed to produce large volumes of goods. When excellent performance and durability are needed, pultruded profiles are employed in a variety of applications, such as transportation, infrastructure, and building. Pultrusion, however, can only create components with consistent cross-sections; it is unable to create complicated shapes.

5.3.8 Resin transfer molding (RTM)

Large and complex composite structures can be produced using the closed-mold method known as RTM. It entails pressurized injection of the fiber-matrix mixture into a closed-mold cavity. High-performance green composites may be made with this technique, which provides exceptional control over the distribution of fibers. In order to demonstrate the better mechanical capabilities and thermal stability of sisal-reinforced PLA composites, Samouh et al. (2019) looked into the application of RTM in this process. The first step in the procedure is to fill a mold cavity with dried natural fibers. Next, under pressure, the biopolymer resin is poured into the mold, filling the mold cavity and saturating the fibers. The resin is cured by heating the mold, and as it cools, the resin solidifies into a composite substance. Among the many benefits of RTM is its capacity to create huge, intricate parts with a high fiber content and superior mechanical qualities. Accurate control over fiber orientation and distribution is also made possible by this approach, resulting in composites with uniform characteristics. To further enhance customization of green composites to meet particular application requirements, RTM is a versatile method that may integrate many kinds of biopolymers and natural fibers.

5.3.9 Vacuum infusion

An innovative manufacturing technique called vacuum infusion is utilized to create composite constructions that are flawless and devoid of voids. Dry reinforcing fibers are inserted into a mold, and then, under vacuum pressure, resin is infused into them. Vacuum infusion can be used to create natural fiber-reinforced epoxy composites, as shown in a study by Patel et al. (2023), which highlights the process's capacity to create lightweight, highly strength components with few voids. First things first: get the mold ready and place the reinforcements made of dry fiber. Air is then drawn out of the mold cavity by applying a vacuum while a vacuum bag is fastened over the mold. In order to ensure complete wetting and consolidation, resin is injected into the mold through an inlet and drawn through the fiber reinforcements by vacuum pressure. A robust, void-free composite structure is produced by allowing the laminate to cure after the resin has completely saturated the fibers. Many benefits come with vacuum infusion, one of which is its capacity to create superior composites with minimum voids and outstanding mechanical qualities. As a consequence, components with consistent and high performance can be produced by precisely controlling the resin content and fiber volume fraction through this technique. In comparison to conventional open-mold methods, vacuum infusion is a more ecologically friendly process because it also minimizes the emission of VOCs.

5.3.10 Additive manufacturing

Additive manufacturing, commonly referred to as 3D printing, is a swiftly advancing technology that presents novel opportunities for fabricating green composites. This technology provides flexibility in producing complex geometries and custom designs, expanding application domains for green composites. The process begins with the preparation of a fiber-matrix mixture, which is fed into a 3D printer. The printer then deposits the mixture layer by layer, till the composite material forms into a desired shape. Celik et al. (2025) investigated the use of Additive manufacturing/3D printing to fabricate hemp-reinforced PLA composites, highlighting their potential for customized packaging and consumer products. A schematic diagram of additive manufacturing technologies is shown in Figure 5.5. This process is highly automated and can produce components with details and tight tolerances. The precise control of processing parameters, such as layer thickness, deposition rate, and curing temperature, is a major challenge. These parameters can significantly impact the quality and performance of the final product. Despite these challenges, additive manufacturing remains a promising

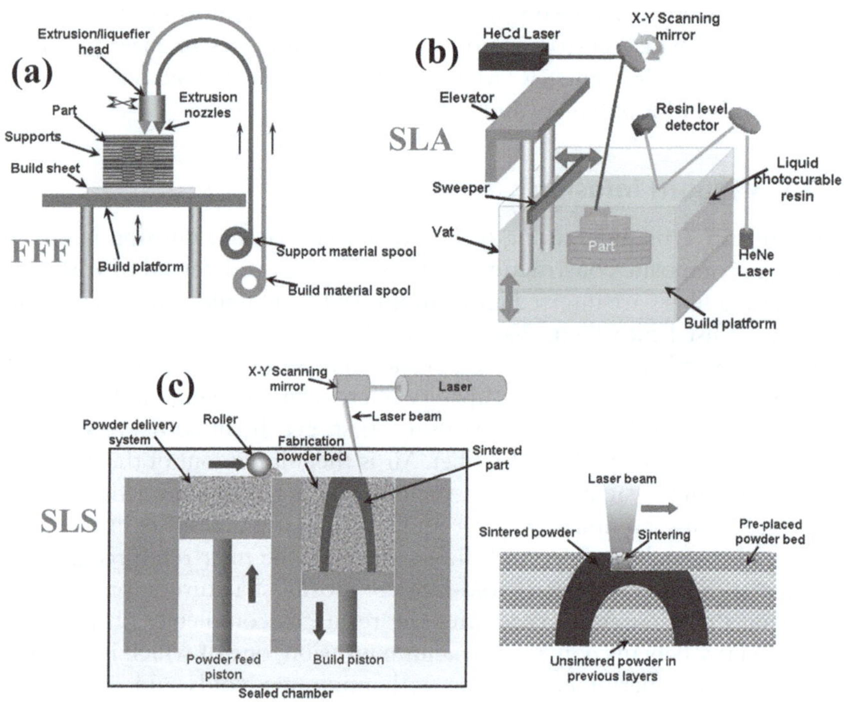

Figure 5.5 Additive Manufacturing technologies (a) FFF, (b) SLA, and (c) SLS.

Table 5.3 Additive manufacturing (AM) methods for fabricating composites

Method	*Description*
Material extrusion	Molten material is extruded through a heated nozzle onto a build plate. Known as fused deposition modeling (FDM) or fused filament fabrication (FFF), this technique is recognized for its low cost, minimal waste, and capability to create intricate geometries.
Vat photopolymerization	This method uses UV-cured photopolymers for fabrication, commonly referred to as stereolithography (SLA). While it provides excellent build quality, the range of photopolymer materials is limited.
Powder bed fusion	This method, known as selective laser sintering (SLS), utilizes thermal energy to fuse a powder bed.

method for the production of green composites due to its ability to produce complex and customized components with high precision. As outlined by ASTM International Technical Committee F42, several additive manufacturing (AM) techniques for manufacturing composites are detailed in Table 5.3.

FFF is the most commonly employed technique for biocomposites, offering advantages such as low cost, efficiency, and the capability to produce complex shapes. Polymer matrix composites with fillers and fibers are produced using AM techniques such as FFF and SLS. These composites show enhanced properties, though nanoscale fillers can lead to agglomeration issues. SLA has been used to develop multimaterial structures with properties tailored to specific sites. Incorporating nanofillers, surface treatments, and fiber reinforcements can improve SLA feedstock properties, though challenges like UV light inhibition remain. The FFF remains the most used technique for fabricating composite materials.

5.4 APPLICATIONS OF GREEN COMPOSITES

Green composites have found applications in diverse industries, including automotive, construction, packaging, consumer goods, aerospace, and medical devices. This section will discuss the various applications of green composites, highlighting recent advancements and their potential to replace conventional materials.

5.4.1 Automotive industry

Green composites have gained popularity in the automobile industry because of their excellent mechanical qualities and lightweight design. Because they may cut weight significantly, these materials are being utilized more often in vehicle components, which improves fuel efficiency and lowers greenhouse gas emissions. Green composites are used in the interior component space to fabricate dashboards, seat constructions, and panels. The usage of bamboo-reinforced green composites for automotive interior components was investigated by X. Wang et al. (2020), showing how they may replace traditional materials and lighten cars. Natural fibers are included in car interiors to improve the vehicle's overall aesthetic and tactile qualities as well as sustainability. Green composites offer an environmentally friendly substitute for conventional materials for external parts, including body panels, fenders, and bumpers. The application of flax-reinforced PLA composites for vehicle exterior components was studied by Kumar and Patel (2022), who showed the materials' exceptional mechanical qualities and environmental advantages. Natural fibers are used in external components to improve impact resistance and decrease the vehicle's overall weight, which improves fuel efficiency.

5.4.2 Construction

One further sector where green composites have shown great promise is the construction industry. These materials are perfect for use in construction because of their exceptional strength-to-weight ratio, resilience to environmental deterioration, and longevity. Structural elements, including panels, beams, and columns, are made of green composites. A study conducted in 2000 by Y. Li et al. showed how sisal-reinforced green composites may be used in structural applications, hence reducing carbon footprints and replacing traditional construction materials. One further sector where green composites have shown great promise is the construction industry. These materials are perfect for use in construction because of their exceptional strength-to-weight ratio, resilience to environmental deterioration, and longevity. An environmentally friendly substitute for traditional insulation goods is provided by the use of green composites in insulation materials. The application of jute-reinforced green composites for building insulation was examined in a study by T. Ahmed et al. (2024), highlighting the materials' advantageous thermal and environmental characteristics. Natural fibers improve heat conductivity and lessen their environmental impact when added to insulating materials, which improves indoor air quality and energy efficiency. Decorative pieces such as wall panels, flooring, and furniture also incorporate green composites. Zhang et al.'s (2022) study examined the usage of bamboo-reinforced green composites for ornamental purposes, highlighting the materials' sustainable nature and visual attractiveness. Enhancing tactile and visual appeal, the use of natural fibers in decorative elements improves indoor aesthetics and promotes environmental sustainability.

5.4.3 Packaging

In addition to addressing environmental issues, the development of biodegradable packaging materials offers better performance qualities, such as improved mechanical strength and barrier qualities. Green composites' biodegradability and barrier qualities have drawn a lot of interest to their application in food packaging. The usage of PLA-based green composites for food packaging applications was investigated by Prasad et al. in 2022, and their results showed that these materials have the ability to replace traditional plastics while still having good barrier qualities against gasses and moisture. The use of biodegradable materials in food packaging enhances food safety and quality. Consumer packaging applications for green composites include biodegradable bags, containers, and wraps. Adding natural fibers to consumer packaging improves mechanical properties and reduces environmental impact, which in turn improves product safety and promotes environmental sustainability. Green composites provide a sustainable substitute for traditional materials in industrial packaging, including boxes, pallets, and protective coverings. The exceptional mechanical characteristics and biodegradability of PBS-based green composites were showcased by Kim et al. (2021) in their demonstration of their usage for industrial packaging. The supply chain's overall sustainability is increased and waste is decreased when biodegradable materials are used in industrial packaging.

5.4.4 Aerospace

Green composites have garnered interest from the aerospace industry due to their better mechanical qualities and lightweight nature. Green composites may significantly reduce weight in aeronautical components, improving fuel efficiency and reducing greenhouse gas emissions. Fuselage panels, wings, and interior components are examples of structural parts made of green composites. By demonstrating the application of sisal-reinforced green composites for aerospace structural components Sahu and Gupta (2017) and Mohanty et al. (2005) highlighted the materials' potential to replace traditional materials and minimize aircraft weight. By improving mechanical qualities and lowering total weight, the use of natural fibers in aircraft components improves environmental sustainability and fuel efficiency. Green composites are also used in the fabrication of interior components, such as seats, cabin panels, and storage compartments. Kumar et al. (2024) investigated the use of flax-reinforced PLA composites for aerospace interior components, demonstrating their superior mechanical properties and environmental benefits. The incorporation of natural fibers in interior components enhances the strength and durability while reducing the overall weight, contributing to improved passenger comfort and sustainability.

5.4.5 Biomedical

Green composites are biocompatible and biodegradable, they may be used for a variety of medical purposes, and this has led to considerable improvements in the biomedical area. Green composites are used to make screws, plates, and scaffolds in the field of surgical implants. The application of PHA-based green composites for surgical implants was studied by Chen et al. (2023), who showed that these materials have the capacity to regenerate tissue and are biocompatible. Medical treatments are safer and more successful overall when biodegradable materials are used in surgical implants because they lower the chance of long-term problems. Green composites provide a sustainable solution for tissue engineering. C. Zhang et al. (2015) explored the use of PLA-based green composites for tissue engineering scaffolds, highlighting their ability to support cell growth and tissue regeneration. The incorporation of natural fibers into tissue engineering scaffolds improves the mechanical properties and biocompatibility, which contributes to better patient outcomes and promotes environmental sustainability. They are also used in delivery systems, providing controlled release and biodegradability. Research by Gigli et al. (2016) demonstrated the use of PBS-based green composites for drug delivery applications, showcasing their potential to enhance patient outcomes and reduce environmental impact.

5.4.6 Machine learning and artificial neural networks (ANN) integration

Integrating ML and ANN opens up new avenues for enhancing the properties and performance of green composites. ML algorithms can process extensive datasets and detect patterns, enabling the prediction and optimization of material properties.

Property Prediction: ML algorithms can predict the mechanical, thermal, and biodegradability properties of green composites based on their composition and processing parameters. Research by Y. Wang et al. (2024) demonstrated the use of ML algorithms to predict the tensile strength and thermal stability of green composites, highlighting their potential to optimize material properties and reduce the need for extensive experimental testing. An example of an ANN model for predicting the properties of a composite material is shown in Figure 5.6.

Process Optimization: ML algorithms can optimize the processing parameters of green composites, such as temperature, pressure, and cooling rate, to improve their quality and performance. A study by Patel and Singh (2023) investigated the use of ML algorithms to optimize the injection molding process of green composites, demonstrating their potential to enhance product quality and reduce production costs.

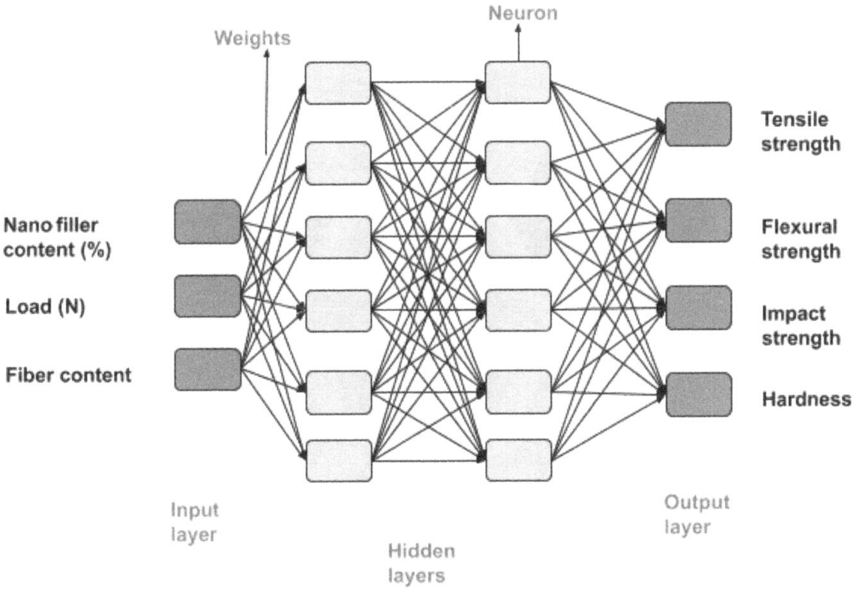

Figure 5.6 An example of an ANN model for predicting the properties of a composite material.

Material Design: ML algorithms can assist in the design of new green composites with tailored properties for specific applications. Research by W. Wang et al. (2023) explored the use of ML algorithms to design hybrid green composites with enhanced mechanical and thermal properties, showcasing their potential to create customized materials for advanced applications.

Failure Analysis: ML algorithms can examine the failure mechanisms of green composites and forecast their lifespan and performance under various conditions. Kumar et al. (2022) illustrated the use of ML algorithms to analyze the failure modes of green composites, underscoring their potential to enhance material durability and reliability.

5.5 FUTURE TRENDS AND CHALLENGES

The continued advancement of novel biodegradable materials and sustainable processing techniques will be crucial in driving further progress in the field of green composites. Ongoing research efforts are focused on developing new materials, optimizing processing methods, and exploring innovative applications. The future trends of green composites are shown

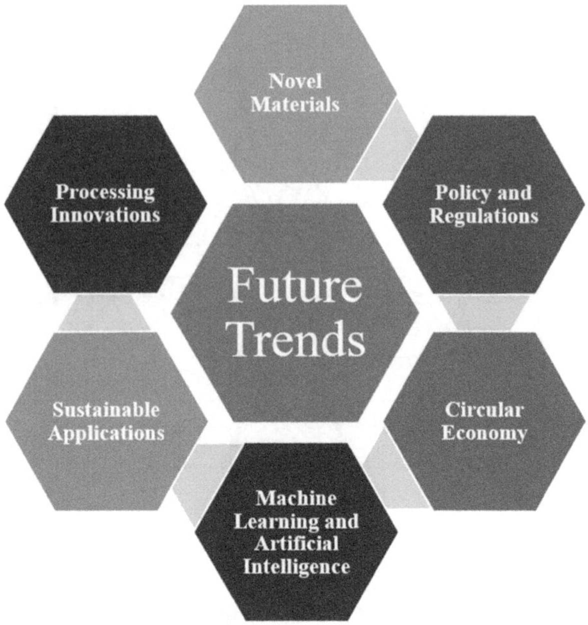

Figure 5.7 Figure showing future trends of green composites.

in Figure 5.7. The future trends of green composites research are going in the direction of working on the development of new novel materials such as hybrid composites and functionalized fiber. Advanced processing innovations such as additive manufacturing of nanocomposites are rapidly progressing. The integration of green composites into various industries such as packaging, automotive, construction, and biomedical underscores their potential to contribute to a more sustainable future. Also, the combination of ML and artificial intelligence offers new opportunities for optimizing the properties and performance of green composites. ML algorithms can analyze extensive datasets and detect patterns, enabling the prediction and enhancement of material characteristics. The creation of green composites aligns with the principles of the circular economy, which focuses on reducing waste and maximizing resource efficiency. Ongoing research efforts focus on developing recycling methods and end-of-life management strategies to ensure the sustainability of green composites. They also focus on developing regulatory frameworks that incentivize the use of sustainable materials and support the commercialization of green composites.

The dynamic landscape of green composites presents both exciting opportunities and challenges. This section will discuss the challenges that need to be addressed to ensure their widespread adoption and long-term viability. These challenges include optimization of fiber-matrix interactions, scalability of production, cost-effectiveness, and environmental impact.

The compatibility between natural fibers and biopolymers is crucial in determining the overall performance of green composites. Research efforts continuously focus on enhancing fiber-matrix interactions through surface treatment methods and functionalized fibers. Additionally, the scalability of production poses a key challenge in the commercialization of green composites. Efficient and cost-effective manufacturing processes are essential to meet the growing demand. Cost-effectiveness is another critical factor influencing the widespread adoption of green composites. Research efforts aim to optimize production processes and reduce raw material costs. The environmental impact of green composites is a key consideration, with lifecycle assessment (LCA) being used to evaluate their sustainability, covering raw material extraction, production, use, and end-of-life disposal. Standardization and regulations are essential to ensure the quality and performance of green composites.

5.6 CONCLUSION

The field of green composites has experienced substantial growth, fueled by the increasing global demand for sustainable and eco-friendly materials. This chapter has detailed the advancements in the composition, processing techniques, and diverse applications of green composites, highlighting their potential to contribute significantly to a sustainable future. The incorporation of natural fibers such as flax, hemp, and jute with biodegradable matrices like PLA and PHAs has improved the mechanical properties and environmental compatibility of these materials. Advanced processing techniques, including injection molding, Compression Molding, RTM, Extrusion, and AM, have enabled the production of high-quality green composites across various industries such as automotive, construction, packaging, aerospace, and biomedical fields. Despite these advancements, challenges such as economic viability, processing consistency, and long-term durability persist. Future research is expected to focus on material innovation, optimization of processing technologies, sustainability metrics, and expanding applications. The integration of ML and ANN is anticipated to play a critical role in optimizing material properties and minimizing the need for extensive experimental testing. Overall, green composites offer significant promise for a sustainable future, with ongoing research set to address current challenges and unlock new opportunities.

REFERENCES

Ahmed, T., Toki, G. F. I., Mia, R., Faridul Hasan, K. M., & Alpár, T. (2024). Mechanical and thermal properties of plant/plant fiber based woven fabric hybrid composites. In S. M. Rangappa, V. Ayyappan, J. Tengsuthiwat, & S. Siengchin (Eds.),

Innovations in woven and non-woven fabrics based laminated composites (pp. 77–113). Springer Nature. https://doi.org/10.1007/978-981-97-7937-6_4
Akil, H. M., Omar, M. F., Mazuki, A. M., Safiee, S. Z. A. M., Ishak, Z. M., & Bakar, A. A. (2011). Kenaf fiber reinforced composites: A review. *Materials & Design*, *32*(8–9), 4107–4121.
Ali, M., Chouw, N., & Patnaikuni, I. (2012). Experimental investigations on coconut-fibre rope tensile strength and pullout from coconut fibre reinforced concrete. *Construction and Building Materials*, *28*(1), 189–199.
Andrew, J. J., & Dhakal, H. N. (2022). Sustainable biobased composites for advanced applications: Recent trends and future opportunities – A critical review. *Composites Part C: Open Access*, 7, 100220.
Antony, S., Cherouat, A., & Montay, G. (2021). Effect of fibre content on the mechanical properties of hemp fibre woven fabrics/polypropylene composite laminates. *Polymers and Polymer Composites*, *29*(9_suppl), S790–S802.
Avérous, L., & Pollet, E. (2012). *Biodegradable polymers*. Springer Science & Business Media.
Bajpai, P. K., Singh, I., & Madaan, J. (2021). Flax fiber-reinforced composites: Mechanical properties and biodegradability. *Journal of Composite Materials*, *55*(2), 189–202.
Barouni, A., Lupton, C., Jiang, C., Saifullah, A., Giasin, K., Zhang, Z., & Dhakal, H. N. (2022). Investigation into the fatigue properties of flax fibre epoxy composites and hybrid composites based on flax and glass fibres. *Composite Structures*, *281*, 115046. https://doi.org/10.1016/j.compstruct.2021.115046
Celik, E., Uysal, M., Gumus, O. Y., & Tasdemir, C. (2025). 3D-printed biocomposites from hemp fibers reinforced polylactic acid: Thermal, morphology, and mechanical performance. *BioResources*, *20*(1), 331–356.
Chen, G., Qiu, X., & Wang, L. (2023). Polyhydroxyalkanoates in green composites for biomedical applications. *Journal of Biodegradable Materials*, *14*(4), 223–234.
Dahal, R. K., Acharya, B., & Dutta, A. (2022). Mechanical, thermal, and acoustic properties of hemp and biocomposite materials: A review. *Journal of Composites Science*, *6*(12), 373. https://doi.org/10.3390/jcs6120373
Dash, M. K., Patro, S. K., & Rath, A. K. (2015). Sustainable use of industrial-waste as partial replacement of fine aggregate for preparation of concrete: A review. *International Journal of Sustainable Built Environment*, *4*(2), 421–439.
Datta, J., & Kopczyńska, P. (2015). Effect of kenaf fibre modification on morphology and mechanical properties of thermoplastic polyurethane materials. *Industrial Crops and Products*, *74*, 566–576. https://doi.org/10.1016/j.indcrop.2015.05.080
El Hadrami, A., Adam, L. R., El Hadrami, I., & Daayf, F. (2010). Chitosan in plant protection. *Marine Drugs*, *8*(4), 968–987.
Felt, O., Buri, P., & Gurny, R. (1998). Chitosan: A unique polysaccharide for drug delivery. *Drug Development and Industrial Pharmacy*, *24*(11), 979–993.
Gigli, M., Fabbri, M., Lotti, N., Gamberini, R., Rimini, B., & Munari, A. (2016). Poly (butylene succinate)-based polyesters for biomedical applications: A review. *European Polymer Journal*, *75*, 431–460. https://doi.org/10.1016/j.eurpolymj.2016.01.016
Guo, R., Xian, G., Li, C., & Hong, B. (2022). Effect of fiber hybrid mode on the tension–tension fatigue performance for the pultruded carbon/glass fiber reinforced polymer composite rod. *Engineering Fracture Mechanics*, *260*, 108208. https://doi.org/10.1016/j.engfracmech.2021.108208

Ibrahim, I. D., Jamiru, T., Sadiku, E. R., Kupolati, W. K., Agwuncha, S. C., & Ekundayo, G. (2015). Mechanical properties of sisal fibre-reinforced polymer composites: A review. *Composite Interfaces*, *23*(1), 15–36.

Jayakumar, R., Prabaharan, M., Nair, S. V., Tokura, S., Tamura, H., & Selvamurugan, N. (2010). Novel chitin and chitosan nanofibers in biomedical applications. *Biotechnology Advances*, *28*(1), 142–150.

Joshi, G., Sharma, V., Saxena, R., & Yadav, K. S. (2021). 17—Polylactic coglycolic acid (PLGA)-based green materials for drug delivery. In S. Ahmed (Ed.), *Applications of advanced green materials* (pp. 425–440). Woodhead Publishing. https://doi.org/10.1016/B978-0-12-820484-9.00017-9

Kim, H., Shin, M. S., Jeon, H., Koo, J. M., Eom, Y., Choi, S., Shin, G., Oh, D. X., Hwang, S. Y., & Park, J. (2021). Highly reinforced poly(butylene succinate) nanocomposites prepared from chitosan nanowhiskers by in-situ polymerization. *International Journal of Biological Macromolecules*, *173*, 128–135. https://doi.org/10.1016/j.ijbiomac.2021.01.102

Kumar, A., Arora, H. C., Kumar, K., Mohammed, M. A., Majumdar, A., Khamaksorn, A., & Thinnukool, O. (2022). Prediction of FRCM–Concrete bond strength with machine learning approach. *Sustainability*, *14*(2), 845. https://doi.org/10.3390/su14020845

Kumar, R., & Patel, T. (2022). Lightweight green composites for automotive applications. *Journal of Automotive Engineering*, *13*(2), 75–88.

Kumar, S., Saha, A., & Zindani, D. (2024). Agro-waste-based polymeric composite laminates for aerospace cabin interior and identification of their optimal configuration. *Biomass Conversion and Biorefinery*, *14*(24), 31907–31923.

Kurose, T., Takemura, A., & Hattori, M. (2014). Development and application of biodegradable soy protein-based resins in packaging materials. *Journal of Polymers and the Environment*, *22*(2), 171–177.

Li, H., Tang, R., Dai, J., Wang, Z., Meng, S., Zhang, X., & Cheng, F. (2022). Recent progress in flax fiber-based functional composites. *Advanced Fiber Materials*, *4*(2), 171–184. https://doi.org/10.1007/s42765-021-00115-6

Li, X., Lin, Y., Liu, M., Meng, L., & Li, C. (2023). A review of research and application of polylactic acid composites. *Journal of Applied Polymer Science*, *140*(7), e53477. https://doi.org/10.1002/app.53477

Li, Y., Mai, Y. W., & Ye, L. (2000). Sisal fibre and its composites: A review of recent developments. *Composites Science and Technology*, *60*(11), 2037–2055.

Maiti, S., Islam, M. R., Uddin, M. A., Afroj, S., Eichhorn, S. J., & Karim, N. (2022). Sustainable fiber-reinforced composites: A review. *Advanced Sustainable Systems*, *6*, 2200258.

Martin, S. (2024). Designing composite-based cylindrical structures and manufacturing composite prototypes by filament winding method.

Mishra, S., Naik, J. B., & Patil, Y. P. (2008). The compatibilizing effect of maleic anhydride on swelling, mechanical, and thermal properties of coir–polypropylene biocomposites. *Journal of Thermoplastic Composite Materials*, *21*(2), 131–147.

Mochane, M. J., Magagula, S. I., Sefadi, J. S., & Mokhena, T. C. (2021). A review on green composites based on natural fiber reinforced polybutylene succinate (PBS). *Polymers*, *13*(8), 1200. https://doi.org/10.3390/polym13081200

Mohanty, A. K., Misra, M., & Drzal, L. T. (2005). *Natural fibers, biopolymers, and biocomposites*. CRC Press. https://books.google.co.in/books?id=AwXugfY2oc4C

Mulenga, T. K., Ude, A. U., & Vivekanandhan, C. (2021). Techniques for modelling and optimizing the mechanical properties of natural fiber composites: A review. *Fibers*, *9*(1). https://doi.org/10.3390/fib9010006

Mungroo, R., Prakash, S., & Khan, W. (2014). Sustainable soy protein-based resins for automotive applications. *Polymer Composites*, *35*(2), 337–344.

No, H. K., Park, N. Y., Lee, S. H., & Meyers, S. P. (2007). Antimicrobial activity of chitosans and chitosan oligomers with different molecular weights. *International Journal of Food Microbiology*, *74*(1–2), 65–72.

Patel, R., & Singh, A. (2023). 3D printing of green composites: Advancements and applications. *Additive Manufacturing Technologies*, *11*(2), 98–110.

Patel, R. V., Yadav, A., & Winczek, J. (2023). Physical, mechanical, and thermal properties of natural fiber-reinforced epoxy composites for construction and automotive applications. *Applied Sciences*, *13*(8), 5126. https://doi.org/10.3390/app13085126

Prasad, A., Kumar, A., & Gajrani, K. (2022). *Biodegradable composites for packaging application*. https://doi.org/10.1201/9781003227908

Rajulapati, V., Dhillon, A., Kumar Gali, K., Katiyar, V., & Goyal, A. (2020). Green bioprocess of degumming of jute fibers and bioscouring of cotton fabric by recombinant pectin methylesterase and pectate lyases from Clostridium thermocellum. *Process Biochemistry*, *92*, 93–104. https://doi.org/10.1016/j.procbio.2020.02.024

Rinaudo, M. (2006). Chitin and chitosan: Properties and applications. *Progress in Polymer Science*, *31*(7), 603–632.

Rowell, R. M., & Rowell, J. (1996). *Paper and composites from agro-based resources*. CRC Press.

Sahu, P., & Gupta, M. (2017). Sisal (Agave sisalana) fibre and its polymer-based composites: A review on current developments. *Journal of Reinforced Plastics and Composites*, *36*(24), 1759–1780. https://doi.org/10.1177/0731684417725584

Samouh, Z., Molnar, K., Boussu, F., Cherkaoui, O., & El Moznine, R. (2019). Mechanical and thermal characterization of sisal fiber reinforced polylactic acid composites. *Polymers for Advanced Technologies*, *30*(3), 529–537. https://doi.org/10.1002/pat.4488

Singh, M. M., Kumar, H., Kumar, G. H., Sivaiah, P., Nagesha, K. V., Ajay, K. M., & Vijaya, G. (2020). Determination of Strength Parameters of Glass Fibers Reinforced Composites for Engineering Applications. *Silicon*, *12*(1), 1–11. https://doi.org/10.1007/s12633-019-0078-3

Tutek, K., & Masek, A. (2022). Hemp and its derivatives as a universal industrial raw material (with particular emphasis on the polymer industry)—A review. *Materials*, *15*(7), 2565. https://doi.org/10.3390/ma15072565

Wang, W., Wu, Y., Liu, W., Fu, T., Qiu, R., & Wu, S. (2023). Tensile performance mechanism for bamboo fiber-reinforced, palm oil-based resin bio-composites using finite element simulation and machine learning. *Polymers*, *15*(12), 2633. https://doi.org/10.3390/polym15122633

Wang, X., Yuan, Z., Zhan, X., Li, Y., Li, M., Shen, L., Cheng, D., Li, Y., & Xu, B. (2020). Multi-scale characterization of the thermal – mechanically isolated bamboo fiber bundles and its potential application on engineered composites. *Construction and Building Materials*, *262*, 120866. https://doi.org/10.1016/j.conbuildmat.2020.120866

Wang, Y., Wang, K., & Zhang, C. (2024). Applications of artificial intelligence/machine learning to high-performance composites. *Composites Part B: Engineering*, *285*, 111740. https://doi.org/10.1016/j.compositesb.2024.111740

Zhang, C., Salick, M. R., Cordie, T. M., Ellingham, T., Dan, Y., & Turng, L.-S. (2015). Incorporation of poly(ethylene glycol) grafted cellulose nanocrystals in poly(lactic acid) electrospun nanocomposite fibers as potential scaffolds for bone tissue engineering. *Materials Science and Engineering: C*, *49*, 463–471. https://doi.org/10.1016/j.msec.2015.01.024

Zhang, J., Huang, F., Wu, Y., Fu, T., Huang, B., Liu, W., & Qiu, R. (2022). Mechanical properties and interface improvement of bamboo cellulose nanofibers reinforced autoclaved aerated concrete. *Cement and Concrete Composites*, *134*, 104760. https://doi.org/10.1016/j.cemconcomp.2022.104760

Zimniewska, M. (2022). Hemp fibre properties and processing target textile: A review. *Materials*, *15*(5). https://doi.org/10.3390/ma15051901

Chapter 6

Innovative uses of additive manufacturing for high-performance composites in the automotive and aerospace industries

S. Senthil Kumar, S. Rathinavel, and T. S. Senthilkumar

6.1 INTRODUCTION

6.1.1 Background and significance

The ability of AM to fabricate complex, high-performance parts more quickly and cheaply has a disruptive effect in many industries. Additive manufacturing (AM) especially helped industries such as the automotive and aerospace industries.

Weight reduction in the aerospace industry is very important [1] since lighter aircraft requires less fuel, which would lower the cost of operation and reduce the environmental effect. Aerospace products manufactured using AM include robust, lightweight parts—air ducts, structural brackets, or turbine blades. Such components that have complex geometries to balance weight and strength cannot be built using the conventional modes of production. Finally, through AM, aerospace engineers can test and refine designs more rapidly, hence shortening the development cycle and striking at creativity.

Similarly, AM serves the automotive industry in making high-performance, lightweight component parts such as exhaust systems, engine parts, and customized interior trims [2]. This technology offers prototyping with the required speed, hence enabling carmakers to innovate faster and, therefore, put new models in the market faster. Again, this capacity of being able to affordably make small-batch and custom parts is extremely useful for the car industry since there is a huge demand for custom and limited models

6.1.2 Purpose and scope

AM integrates with composite materials for application in auto and aerospace sectors. This includes an in-depth analysis of the available technologies, materials, procedures, and applications. It allows a review of the latest developments in AM processes on glass fiber composites, carbon fiber-reinforced polymers (CFRPs), stereolithography, fused deposition modeling, and selective laser sintering. Procedures for using AM with composites are

 DOI: 10.1201/9781003527275-6

also discussed from digital modeling through to layering, material deposition, and post-processing. Case studies and practical applications on how these technologies are being applied in producing lightweight, highly functional aerospace and automotive components are also provided.

The review also explains the advantages, challenges, and possible applications of AM in various industries. Some benefits include more material economy, lightweight, more design freedom, and the possibility to fabricate complex geometries, which bring cost reduction and enhanced performance [3]. It addresses the issues of scalability, high materials and equipment costs, the need for harmonized material certifications, and technical barriers related to the long-term behavior of printed parts. Future prospects are discussed in light of new applications, foreseen progress in both AM and composite materials, technology under development, and research in process to overcome current limitations. The following is a detailed evaluation in an attempt to provide useful insights that shall help researchers, engineers, and business professionals to make use of additive manufacturing and composite materials and improve both processes and products.

6.2 ADDITIVE MANUFACTURING (AM) AND COMPOSITE MATERIALS

6.2.1 The evolution of AM technology

In the 1980s, stereolithography opened up the exponential development of technology in the field of additive manufacturing (AM) [4]. With the help of UV light, which is implemented to cure photopolymer resins layer by layer, this groundbreaking method sets up the current 3D printing production of precise and complicated things from digital models. A number of other AM methods have emerged in those years, all contributing to the further development of science. Developed in the late 1980s, selective laser sintering uses a laser beam to sinter powdered materials, allowing for the production of products with complex shapes and high mechanical strength [5]. One such accessible and popular method is the 3D printing technique that involves fused deposition modeling, dating back to the early 1990s, whereby an object is produced layer upon layer through the extrusion of large numbers of thermoplastic filaments past the heated nozzle [6]. Electron Beam Melting (EBM), developed since 1990 as a technique that produces high-density metal parts, should have better material qualities when compared with its 1990s counterpart because it can melt metal powders using an electron beam [7]. These advances in AM processes have greatly increased the capability to create complex geometries with tight tolerances and repeatability, transforming the manufacturing process in many different industries.

6.2.2 Properties and advantages of composite materials

One of the most representative examples of composite materials are CFRPs. They represent something very near to ideal material engineering since they put together mechanical properties that, until then, had remained unachieved with an actually extraordinary strength-to-weight ratio. These materials play a fundamental role in aerospace and automotive due to the fact that every gram that was saved equated to a gain in performance. On a list of their intrinsic qualities, one counts increased rigidity and corrosion protection and the ability to be tailored exactly to meet performance needs.

6.3 STATE-OF-THE-ART IN AM OF COMPOSITES

6.3.1 Key AM processes

Stereolithography (SLA)
Stereolithography (SLA)—a process whereby solid objects are created layer upon layer from a vat of liquid photopolymer resin, previously cured by ultraviolet radiation. This is done through the selective solidification of resin through a vat of liquid photopolymer resin by means of a UV laser, based on the cross-section of the 3D model that progressively forms the desired shape. SLA finds widespread applications in prototyping and making elaborate models, for it is known to produce models having complex forms in high resolution and with smooth surface finishes.

Selective laser sintering (SLS)
Selective laser sintering (SLS) is an additive process for 3D printing wherein powdered materials are consolidated to a solid structure under a powerful laser, such as metals, polymers, or ceramics. There is no need for support structures in this process, and the powder material will fuse only at points specified by a digital model, layer after layer. SLS allows processing of a diversity of materials and producing parts that have the best mechanical properties with complex geometry.

Fused deposition modeling (FDM)
In FDM, thermoplastic filament is melted and extruded onto a build platform through a heated nozzle. While it is being extruded, the material cools and subsequently solidifies, creating an object layer by layer. It is known that FDM has a big advantage in terms of ease of use, cost, and number of material choices. It is usually used to make working parts with medium mechanical properties, low-volume production, and rapid prototyping.

Electron beam melting

In EBM, an electron beam selectively melts metal powder in a high-vacuum environment to create dense metal parts layer by layer [8]. As such, EBM works perfectly in creating complex metal parts with maximum durability and high material qualities. It can create parts of complex geometries with excellent mechanical performance that have great applications in aerospace, medical, and automotive industries for functional metal part productions.

Direct metal laser sintering (DMLS)

Another technology, very similar to SLS but dedicated to metal powders, is DMLS—Direct Metal Laser Sintering. It is also known as Direct Metal Laser Melting or DMLM [9]. The technology creates fully dense metal objects through the selective fusing of metal powders layer by layer using a high-power laser. Notably, DMLS finds applications in the production of complex, highly accurate metal parts with prime mechanical properties, thus quite suitable for automotive, aerospace, and medical industries.

Material jetting

Material jetting is like inkjet printing; droplets of a photopolymer or wax-based material are deposited on a build platform instead of ink. The droplets are then cured or cemented layer by layer through heat or UV radiation to build the desired object. With material jetting, it's possible to have parts with multiple colors or materials, complex features, and very fine accuracy. It is also used in making complicated models or molds, for dental applications, and for rapid prototyping [10].

Figure 6.1 shows a simple flowchart giving the classification of AM processes.

Table 6.1 provides an overview of various AM techniques employed nowadays.

6.3.2 Composite materials employed

Carbon fiber-reinforced polymers (CFRPs)

Composite materials such as CFRPs are defined as having carbon fibers embedded in a polymer matrix material, with the matrix consisting of epoxy resin. These are known for their highly developed stiffness and corrosion resistance along with a strength-to-weight ratio [11]. They have found extensive application in those sectors such as aerospace, automotive, and sports goods, having a need for lightweight and high-performance parts, for example, in aircraft fuselages, automobile body parts, and bicycle frames.

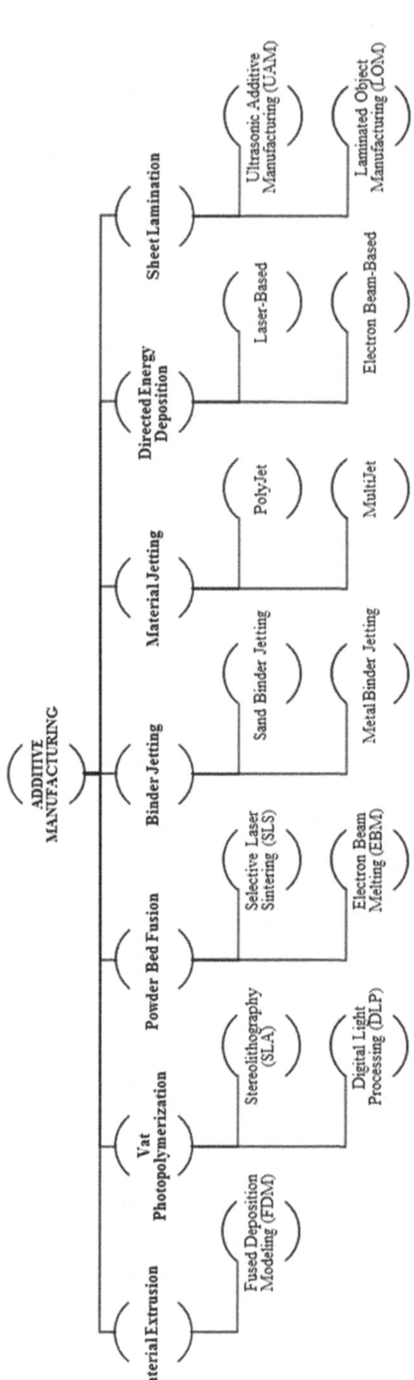

Figure 6.1 Classification of additive manufacturing processes.

Table 6.1 Overview of additive manufacturing techniques

Sl. No	*Technique*	*Material type*	*Process summary*	*Applications*
1.	Stereolithography (SLA)	Photopolymer resin	Cures the liquid resin using a UV laser	Prototype modeling, jewelry making, dental models
2.	Selective laser sintering (SLS)	Powdered form of materials	Laser is used to sinter the powdered material, thereby fusion of materials takes place	Aerospace and automotive functional parts
3.	Fused deposition modeling (FDM)	Thermoplastic filament	Extrusion of molten thermoplastic layer by layer	Prototype modeling and consumer products
4.	Electron beam melting	Metal powders	Electron beams are focused on the metal powders	Medical implants, high-strength parts, aerospace parts
5.	Material jetting	Photopolymers	Deposits material in the form of droplets, cured by UV light	Full-color prototypes, dental and medical implants

Glass fiber-reinforced polymers (GFRPs)

Glass fibers are impregnated into a polymer matrix, often polyester or epoxy resin, forming glass fiber-reinforced polymers, or GFRPs. Comparatively, GFRPs are relatively expensive than CFRPs which has high strength, stiffness, and corrosion resistance [12]. They have been applied to a number of parts handling many different applications, such as the hull of boats, automobile body components, and construction elements in buildings and bridges within composite, shipbuilding, and automotive industries.

Kevlar-reinforced polymers (KRP)

Kevlar-Formed Aramid fibers such as Kevlar find their application in polymers and are typically embedded in an epoxy resin polymer matrix. Kevlar fibers exhibit high tensile strength, toughness, abrasion, and impact resistance [13]. They are widely used in aerospace and automotive parts where high impact strength and toughness are needed, and in personal protection items like body armor and helmets.

Metal matrix composites (MMCs)
MMCs are the resultant material when carbon or ceramic fibers of high tensile strength are used to reinforce a metal matrix. Through the use of MMCs, the benefits of combining the high strength and stiffness of the reinforcing material with the ductility and formability of the metal matrix material became possible. They find their use in engine parts, brake rotors, armor plating, and other components where a high strength and wear resistance associated with thermal stability is involved in the aerospace, automotive, and military industries [14].

Ceramic matrix composites (CMCs)
Ceramic Matrix Composites are ceramic fibers included in a ceramic matrix, such as silicon carbide or alumina. Such composites provide high-temperature resistance, thermal stability, and more—to an above-average resistance to wear and corrosion [15]. Being industrially used for exhaust systems in automobiles, in the aerospace industry for making turbine blades, and cutting tools, CMCs are applied in heavy industries.

Composite materials are vital for the production of lightweight, high-performance parts during the development of AM technologies with ultimate mechanical qualities and durability. Figure 6.2 shows a simple flowchart of the various composite materials employed in the AM processes.

6.3.3 Techniques and technologies

Continuous fiber reinforcement
Continuous fiber reinforcement is a method to add continuous fibers, which are typically carbon or glass, into AM. This steadily feeds these fibers into a printing nozzle or embeds them within the matrix material to produce parts with augmented strength, stiffness, and toughness [16]. For creating structural parts where good mechanical performance is required in aerospace, automotive, and marine applications, continuous fiber reinforcement works especially well.

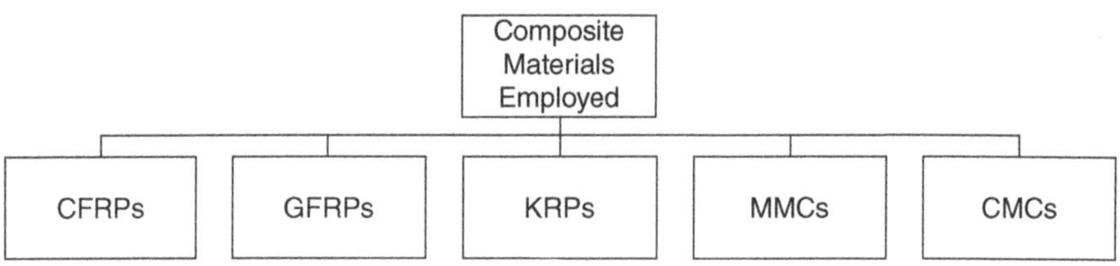

Figure 6.2 Various composite materials employed in the additive manufacturing processes.

Short fiber reinforcement
Short fiber reinforcement is a process for incorporating short fibers into the matrix material using AM. Normally, these fibers could be a few micrometers to several millimeters long. These small strands improve mechanical properties, including strength, stiffness, and impact resistance of the printed parts [17]. With a view to creating components with enhanced performance in different industrial applications, processes of 3D printing, like SLS and fused deposition modeling (FDM), often involve short fiber reinforcement.

Hybrid composites
Hybrid composites incorporate more than one reinforcement material, such as fibers, particles, or flakes, within a single matrix material. This technique allows the variation of the composite's properties to meet specific performance requirements. Hybrid composites can be fabricated through AM processes by combining different reinforcement materials during printing [18]. Such composites offer a tunable solution for applications where a part's strength, stiffness, and impact resistance need to vary within its different areas.

Nanocomposites
In nanocomposites, nanoparticles are usually of sizes between 1 and 100 nanometers that are included into the matrix material to improve, particularly the composite qualities. These nanoparticles can improve the mechanical, thermal, electrical, or barrier properties depending on their composition and dispersion in the matrix. AM techniques can control the distribution and orientation of nanoparticles within the printed objects; therefore, it is possible to create nanocomposites with defined qualities for aerospace industries, electronics, and medicinal applications [19].

Multi-material printing
Multi-material printing enables the printing of complex parts with multiple functions through the concurrent deposition of multiple materials possessing different properties [20]. In this method, several materials—composites, metals, polymers, and ceramics—are combined into one part. AM techniques that selectively deposit components layer by layer realize multi-material printing; examples are Material Jetting and Binder Jetting. Multi-material printing can create objects with complex shapes, with gradient characteristics, or embedded functions, enabling them to find applications in consumer goods, healthcare, and engineering.

It is in materials engineering that AM techniques and technologies for composite materials offer creative solutions for design and fabrication of high-performance components with customized qualities that can satisfy numerous industries.

6.3.4 Material design and innovation

In terms of composite materials, most of the material design innovations relate to their modification and improvement to suit certain uses. Therefore, this necessitates a diversified strategy considering a number of key development areas:

New resin systems
Novel resin systems specifically designed to fulfill the requirements for specific applications are formulated and synthesized as part of material design innovations [21]. Scholars and experts in materials science are continuously researching new polymer chemistries, additives, and curing processes to extend the functionality, strength, and processability of composite materials. These developments open ways to manufacture resins with improved mechanical properties, chemical resistance, and thermal stability, making them more environmentally sustainable, thus further increasing the span of industries where composite materials can find applications.

Enhancing fiber-matrix interactions
Among the most important parameters to be optimized for maximum mechanical properties and general performance of the composite material is the interfacial interaction between the polymer matrix and the reinforcing fibers [22]. It is from this need that the focus of material design innovation in the development of surface treatments, functional coatings, and coupling agents improving fiber adhesion and bonding to the matrix comes. By maximizing the interactions between the fiber and the matrix, problems associated with fiber pull-out, matrix cracking, and delamination can be minimized, hence developing composite materials that exhibit higher stiffness, toughness, and strength.

Utilizing advanced simulation tools
Indeed, with the introduction of advanced simulation tools, this is considered constitutive in the area of innovative material design—tools such as computational modeling and FEA. Those instruments enable researchers to model and predict how composite materials will react to different load scenarios, high temperatures, and environmental impacts [23]. By means of computational methods, researchers can reduce weight, cost, and manufacturing complexity, attaining the targeted performance through the optimization of material compositions, fiber orientations, and processing parameters. Predictive modeling enables faster design of materials, reduces the need for extended testing experimentally, and fosters rapid development of high-performance composite materials tailored for specific applications.

6.4 APPLICATIONS IN AEROSPACE

6.4.1 Lightweight aircraft components

AM enables the fabrication of structurally sound and lightweight parts, as in the case of airframe structures, hinges, and brackets. All of these components are critical to improving fuel efficiency by reducing weight in a variety of industries, especially in aerospace and automotive.

Lightweight components
AM techniques, including FDM and SLS, enable complex internal structures and geometries that, with conventional techniques of production, are difficult or impossible to achieve. This, thus, means it will be easy to make lightweight components without loss of performance or structural integrity. It is through this process that AM enables the production of products with less mass while the mechanical strength and durability can be maintained or improved due to the optimization of geometry and material distribution [2].

Structurally efficient designs
But with AM, one has the benefit of structurally efficient design whereby materials are deliberately located to handle stresses and loads more efficiently [24]. In that respect, conventional production methods may be strongly limited by particular design limitations when optimizing structures for load-bearing efficiency. In sharp contrast, through the layer-by-layer method, AM is able to place materials exactly, thereby allowing the manufacturing of elements with improved structural performance but reduced material consumption, which is associated with better load routes. Hence, the fabricated products are lighter, stronger, and more resistant compared to those realized through traditional methods of production [25].

Contribution to weight reduction and fuel efficiency
Lightweight and structurally sound parts developed through AM mean a great reduction in the weight of cars and airplanes. In aerospace, this means weight reduction in airframe structures, brackets, and hinges, and hence less fuel consumption, increased range, and less pollutants. For the automobile sector, lightweight parts translate into better overall vehicle performance, longer battery life in electric cars, and increased fuel efficiency.

Basically, AM allows for unprecedented design freedom, material efficiency, and performance optimization in changing the face of the manufacturing process—this is because of the production of lightweight, structurally efficient parts. Harnessing these advantages will realize huge ecological, fuel-related, and weight-reducing benefits, which in the long run will drive further innovation and progress in transportation and more.

6.4.2 Advanced satellite systems

In reality, the accuracy and personalization of the technique of AM are unparalleled, making it perfect for setting up intricate parts of satellites, such as structural supports and antenna arrays [26]. Thanks to the peculiar properties of AM, complex geometries and custom-made designs that cannot be easily or are impossible to create by traditional manufacturing techniques can be established.

Precision
AM techniques, such as stereolithography and SLS, provide high precision and accuracy for item details and tight tolerance achievement. Satellite components have very critical accuracy, even in minor deviations in size or alignment, which might affect performance [27]. Using AM's precision will ensure that satellite components will comply with specifications and work well in space conditions.

Customization
AM allows for unprecedented customization in satellite component fabrication compared to traditional manufacturing methods. With this, engineers can go on to make parts that have very precise shapes, sizes, and functionalities, depending on the mission requirement. This improves overall performance and the reliability of the satellite system, as due to this customization, the functionality and performance of components can be optimized. For instance, the shapes of the antenna arrays can be adjusted precisely to ensure strong signal transmission and reception, while structural supports can be lightened without compromising stiffness or strength.

Performance and reliability
AM accuracy and adaptability increase the dependability and performance of satellite systems. With custom-designed components, fault or failure possibilities are reduced since they ensure excellent operation and compatibility with other elements in the system. Moreover, with AM, it is possible to produce large components that are both strong and lightweight; this holds significant interest for satellites because every gram counts. Satellite manufacturers can use AM to enhance system performance, mission success rates, and operational lifetimes.

6.4.3 Space exploration technologies

AM is already crucial for creating advanced space exploration technologies. For example, the construction of rovers and habitats will demand the ability to create highly complex, extremely lightweight, and very strong parts to perform well under harsh extraterrestrial conditions.

Complex components
AM enables the production of complex geometrics and structures that are unattainable or extremely difficult to build using conventional methods of manufacturing. This is extremely beneficial in space-related technologies, where parts often have to be tailored to unique mission specifications or environmental limitations. For example, in the construction of complex robotic parts for rovers with exact functioning, or very complex lattice structures for lightweight but resilient dwelling modules, AM can be used [28].

Lightweight design
All this makes weight of the essence in space missions, where every gram of payload means money in the launch cost and feasibility of the mission. AM, such as FDM and SLS, allows lightweight component production by an optimized use of material, avoiding superfluous mass [29]. With AM, space technologies are able to generate space exploration technologies that are both efficient and cost-effective and can fulfill the mission requirements through weight reduction without compromising strength and functionality [1].

High-strength material
The material for the parts intended for space exploration must be strong and robust to withstand space conditions, including radiation, extreme temperatures, and vacuum. On top of all, parts must be lightweight. AM provides a unique array of materials that includes the metallic alloys and the CFRPs, from which the outstanding mechanical properties can be used to fit the requirements of working materials used in space. The capability of making high-strength materials with ADD has facilitated the creation of solid, reliable space exploration systems, which will be resistant to either the extreme conditions in space during exploration or during flight.

Innovative and flexibility
One of the biggest advantages of AM is its ability to support innovation and flexibility within the design and fabrication process. Engineers or designers can try very many designs in a short period, undertake tests of

their concepts, and effect changes at insignificant costs instead of paying for tooling or setup of manufacture. This fast rate of change actually does help much, particularly in the very dynamic and fast space exploration, where technology drivers and mission requirements quite often change very quickly [30]. With this, space agencies and private companies can take up AM and increase the speed at which innovative space exploration technologies are developed to throw the edges and possibilities open to their space exploration.

6.5 APPLICATIONS IN AUTOMOBILE INDUSTRY

6.5.1 Composite materials in automobile manufacturing

The strategic approach in employing composite materials in vehicle manufacturing explains a continuous quest to reach better superiority and fuel efficiency. The possibility of such weight reduction, meanwhile performing and maintaining structure and durability equivalent to or beyond current practice, is attained by the substitution of traditional materials with composite materials. Here are a few applicable examples of the same in the automotive sector:

Body panels
Body panels such as hoods, doors, fenders, and roofs are now more commonly made from composite materials such as CFRPs and glass fiber-reinforced polymers [31]. Such lightweight panels add to reduced weight in the overall automobile, hence loaning better handling and fuel economy. Moreover, the design freedom of composite body panels additionally gives the automobile manufacturer the opportunity to provide aerodynamic shapes and detailed stylings that all add together to give the vehicle performance as well as aesthetic appeal.

Interior components
Composites are used for the dashboard panel, door panels, and seat frames. The use of these materials in interior automotive components continues to prove helpful in enabling vehicle manufacturers to reduce weight without compromising comfort or functionality [32]. Among the merits of using composite interior components, one notes better thermal and acoustic insulation qualities and a decline in NVH. This very flexibility of application also enables the incorporation of state-of-the-art functionalities, including electronic sensors and integrated airbags, to provide extra protection and comfort for the occupants.

Structural parts
Main automotive structural parts that are manufactured with the aid of composite materials include crash structures, suspension parts, and

chassis reinforcements. By adopting composites for use in structural applications, automakers can save a great amount of weight while maintaining developments in rigidity and crashworthiness, which can also be improved. High strength-to-weight ratios, resistance to corrosion, and flexibility in design of composite structural parts have made it possible to develop vehicles with higher performance and safety. Besides increasing structural performance and efficiency, composites allow integration of innovative materials and processes, like hybrid laminates or personalized fiber placement [33].

6.5.2 AM for automotive components

AM has grown to become one of the enabling technologies for the automotive industry in rapid prototyping and production of complex components. It offers design freedom and manufacturing flexibility, which is unparalleled in the history of manufacturing. Here is how AM can help in the development of automotive components:

Rapid prototyping
Specifically, SLS and SLA represent two AM technologies that attain high-speed and high-accuracy prototyping in automotive components. Without tooling costs or conventional manufacturing methods, designers and engineers are allowed to iterate designs as rapidly as possible, test new ideas, and evaluate performance. In this way, this rapid prototyping approach will increase the speed of development associated with new products and will help in exploring creative design ideas more easily.

Production of complex components
Using AM, complex automobile parts can be fabricated that cannot be done by conventional methods. Lightweight structures, internal channels, and complex geometries can be fabricated by techniques like FDM and Direct Metal Laser Melting, among others. It is possible under this process to create parts of an engine, fully customized interiors, and chassis components designed to meet some performance specifications that look more pleasant in appearance.

Lightweight and optimized structures
AM enables the production of lightweight structures that are optimal in their shape by intentionally adding material only where it is needed, reducing waste and increasing performance. Through advanced design software and optimization techniques, engineers can develop parts that feature optimized load paths, reduced mass, and improved strength-to-weight ratios. Topology optimization and lattice structures are two AM techniques that further maximize lightweighting potential in automotive components for increased vehicle performance, handling, and fuel economy [34].

6.6 CONCLUSION

6.6.1 Summary of main points

The amalgamation between AM and composite materials has achieved a quantum jump into the automotive and aerospace industries, hence thereby advantaging it. Lightweighting: composite parts manufactured via AM lead to lightweight automotive and aircraft structures, which lowers fuel consumption and turns out to increase efficiency. This is big for cost and environmental effectiveness.

Emission reduction: Fulfilling regulations and environmental duty in the aerospace and automobile industries will become possible through reducing emissions by creating lightweight structures that AM technology can enable.

High performance: AM makes it possible to create complex geometries and customized designs for highly performance-rich parts that are incomparable to conventional methods of manufacturing, ensuring enhanced functionality and reliability.

State-of-the-art technologies: It shows the advancement in AM methods, materials, and techniques in the best possible way, showing the willingness of this industry to accept new ways and start fresh approaches to challenging manufacturing problems.

6.6.2 Implications for the aerospace and automobile industries

The evolving process of AM and composite materials is likely to give a new vista entirely to the aerospace and automobile industries and make more innovations and new developments possible.

Efficiency and performance: AM and composites will make production of unbeatable cars and aircraft of the future possible. The lightweight components produced by AM will reduce fuel consumption, increase range, and enhance maneuverability, providing total vehicle and aircraft performance.

Environmental sustainability: AM and composite material development and application will mean major steps toward environmental sustainability. Lightweight structures and optimal designs may potentially reduce emissions, fuel consumption, and environmental effects, thus supporting international efforts on climate change and promoting sustainable transport options.

Future cars and airplanes will be unequaled in their efficiency and performance—quite simply, due to the fact that AM and composites realize this vision. Lightweight components coming out of AM processes will assure the optimal total vehicle and aircraft performances by minimization of fuel consumption, increase of the range, and maximization of maneuverability.

Customization and innovation: Dramatic improvements in environmental sustainability will benefit from the use of AM and composite materials. Light structures and optimized designs will lower emissions, fuel consumption, and environmental effects, thus contributing to international efforts on the fight against climate change and the promotion of sustainable transport means.

Market competitiveness: These businesses will, in turn, gain a clear advantage over all their competitors in the industry, especially on matters concerning AM and composite technologies. This would support manufacturers in reducing time-to-market, which accelerates the speed of product development cycles through lightweight construction, rapid prototyping, and design flexibility, hence showing differentiation in a highly competitive market.

Collaborative partnerships: It is through collaborative efforts among the government, research institutions, and the industry that innovation will be sparked and the adoption of AM and composite materials hastened. Public-private partnerships would be foreseen to foster collaboration in generating breakthroughs that offer solutions across industries through information sharing, technology transfer, and investment in research and development efforts.

REFERENCES

1 Williams, J.C. and Starke, E.A. 2003. Progress in structural materials for aerospace systems. *Acta Materialia*, 51(19), 5775–5799. doi: 10.1016/j.actamat.2003.08.02

2 Gu, D. 2015. *Laser Additive Manufacturing of High-Performance Materials*. Springer eBooks. doi: 10.1007/978-3-662-46089-4

3 Ford, S. and Despeisse, M. 2016. Additive manufacturing and sustainability: An exploratory study of the advantages and challenges. *Journal of Cleaner Production*, 137, 1573–1587. doi: 10.1016/j.jclepro.2016.04.150

4 Jiménez, M., Romero, L., Domínguez, I.A., Espinosa, M. del M. and Domínguez, M. 2019. Additive manufacturing technologies: An overview about 3D printing methods and future prospects. *Complexity*, 2019, 1–30. doi: 10.1155/2019/9656938

5 Spears, T.G. and Gold, S.A. 2016. In-process sensing in selective laser melting (SLM) additive manufacturing. *Integrating Materials and Manufacturing Innovation*, 5(1). doi: 10.1186/s40192-016-0045-4

6 Boparai, K.S. and Singh, R. 2017. Advances in fused deposition modeling. *Reference Module in Materials Science and Materials Engineering*. doi: 10.1016/b978-0-12-803581-8.04166-7

7 Murr, L.E., Gaytan, S.M., Ramirez, D.A., Martinez, E., Hernandez, J., Amato, K.N. and Wicker, R.B. 2012. Metal fabrication by additive manufacturing using laser and electron beam melting technologies. *Journal of Materials Science & Technology*, 28(1), 1–14. doi: 10.1016/s1005-0302(12)60016-4

8 Öchsner, A. and Altenbach, H. (Eds.). 2020. Engineering design Applications III. *Advanced Structured Materials*. doi: 10.1007/978-3-030-39062-4

9 Senthilkumar, T.S., Senthil Kumar, S. and Rathinavel, S. 2023. Performance analysis of conventional and DMLS copper electrode during EDM process in AA4032-TiC composite. *3D Printing and Additive Manufacturing*, 10(3). https://doi.org/10.1089/3dp.2021.0030

10 Dilag, J. Chen, T. Li, S. and Bateman, S.A. 2019. Design and direct additive manufacturing of three-dimensional surface micro-structures using material jetting technologies. *Additive Manufacturing*. doi: 10.1016/j.addma.2019.01.009

11 Pramanik, A. Basak, A.K. Dong, Y. Sarker, P.K. Uddin, M.S. Littlefair, G. and Chattopadhyaya, S. 2017. Joining of carbon fibre reinforced polymer (CFRP) composites and aluminium alloys: A review. *Composites Part A: Applied Science and Manufacturing*, 101, 1–29. doi: 10.1016/j.compositesa.2017.06

12 Nkurunziza, G. Debaiky, A. Cousin, P. and Benmokrane, B. 2005. Durability of GFRP bars: A critical review of the literature. *Progress in Structural Engineering and Materials*, 7(4), 194–209. doi: 10.1002/pse.205

13 Singh, T.J. and Samanta, S. 2015. Characterization of Kevlar fiber and its composites: A review. *Materials Today: Proceedings*, 2(4–5), 1381–1387. doi: 10.1016/j.matpr.2015.07.057

14 Kishawy, H.A. and Hosseini, A. 2019. Metal matrix composites. In: *Machining Difficult-to-Cut Materials. Materials Forming, Machining and Tribology*. Springer, Cham. doi: 10.1007/978-3-319-95966-5_5

15 Krenkel, W. 2005. Carbon fiber reinforced CMC for high-performance structures. *International Journal of Applied Ceramic Technology*, 1(2), 188–200. doi: 10.1111/j.1744-7402.2004.tb00169.x

16 Naranjo-Lozada, J. Ahuett-Garza, H. Orta-Castañón, P. Verbeeten, W.M.H. and Sáiz-González, D. 2019. Tensile properties and failure behavior of chopped and continuous carbon fiber composites produced by additive manufacturing. *Additive Manufacturing*, 26, 227–241. doi: 10.1016/j.addma.2018.12.020

17 Ivey, M. Melenka, G.W. Carey, J.P. and Ayranci, C. 2017. Characterizing short-fiber-reinforced composites produced using additive manufacturing. *Advanced Manufacturing: Polymer & Composites Science*, 3(3), 81–91. doi: 10.1080/20550340.2017.1341125

18 Nunna, S. Chandra, P.R. Shrivastava, S. and Jalan, A. 2012. A review on mechanical behavior of natural fiber-based hybrid composites. *Journal of Reinforced Plastics and Composites*, 31(11), 759–769. doi: 10.1177/0731684412444325

19 Lin, T.C. Cao, C. Sokoluk, M. Jiang, L. Wang, X. Schoenung, J.M. and Li, X. 2019. Aluminum with dispersed nanoparticles by laser additive manufacturing. Nature *Communications*, 10(1). doi: 10.1038/s41467-019-12047-2

20 Singh, R. Kumar, R. Farina, I. Colangelo, F. Feo, L. and Fraternali, F. 2019. Multi-material additive manufacturing of sustainable innovative materials and structures. *Polymers*, 11(1), 62. doi: 10.3390/polym11010062

21 Ilie, N. and Hickel, R. 2011. Resin composite restorative materials. *Australian Dental Journal*, 56, 59–66. doi: 10.1111/j.1834-7819.2010.01296.x

22 Rajak, D. Pagar, D. Menezes, P. and Linul, E. 2019. Fiber-reinforced polymer composites: Manufacturing, properties, and applications. *Polymers*, 11(10), 1667. doi: 10.3390/polym11101667

23 David Müzel, S. Bonhin, E.P. Guimarães, N.M. and Guidi, E.S. 2020. Application of the finite element method in the analysis of composite materials: A review. *Polymers*, 12(4), 818. doi: 10.3390/polym12040818
24 Vaneker, T. Bernard, A. Moroni, G. Gibson, I. and Zhang, Y. 2020. Design for additive manufacturing: Framework and methodology. *CIRP Annals*. doi: 10.1016/j.cirp.2020.05.006
25 Tofail, S.A.M. Koumoulos, E.P. Bandyopadhyay, A. Bose, S. O'Donoghue, L. and Charitidis, C. 2018. Additive manufacturing: Scientific and technological challenges, market uptake and opportunities. *Materials Today*, 21(1), 22–37. doi: 10.1016/j.mattod.2017.07.001
26 Gill, S.S. Arora, H.J. and Sheth, V. 2017. On the development of Antenna feed array for space applications by additive manufacturing technique. *Additive Manufacturing*, 17, 39–46. doi: 10.1016/j.addma.2017.06.010
27 Helena, D. Ramos, A. Varum, T. and Matos, J.N. 2020. Antenna design using modern additive manufacturing technology: A review. *IEEE Access*, 8, 177064–177083. doi: 10.1109/access.2020.3027383
28 Sacco, E. and Moon, S.K. 2019. Additive manufacturing for space: Status and promises. *The International Journal of Advanced Manufacturing Technology*, 105(10), 4123–4146. doi: 10.1007/s00170-019-03786-z
29 Kilian, M., Hartwanger, C., Schneider, M. and Hatzenbichler, M. 2017. Waveguide components for space applications manufactured by additive manufacturing technology. *IET Microwaves, Antennas & Propagation*, 11(14), 1949–1954. doi: 10.1049/iet-map.2016.0984
30 Huang, Y. Leu, M.C. Mazumder, J. and Donmez, A. 2014. Additive manufacturing: Current state, future potential, gaps and needs, and recommendations. *Journal of Manufacturing Science and Engineering*, 137(1), 014001. doi: 10.1115/1.4028725
31 Ahmad, H. Markina, A.A. Porotnikov, M.V. and Ahmad, F. 2020. A review of carbon fiber materials in automotive industry. *IOP Conference Series: Materials Science and Engineering*, 971, 032011. doi: 10.1088/1757-899x/971/3/032011
32 Agarwal, J. Sahoo, S. Mohanty, S. and Nayak, S.K. 2019. Progress of novel techniques for lightweight automobile applications through innovative eco-friendly composite materials: A review. *Journal of Thermoplastic Composite Materials*, 33(7), 978–1013. doi: 10.1177/0892705718815530
33 Elmarakbi, A. (Ed.). 2013. *Advanced Composite Materials for Automotive Applications*. doi: 10.1002/9781118535288
34 Plocher, J. and Panesar, A. 2019. Review on design and structural optimisation in additive manufacturing: Towards next-generation lightweight structures. *Materials & Design*, 108164. doi: 10.1016/j.matdes.2019.108164

Chapter 7

Development of green composite

Conventional and computational testing modeling

Ipoteng Justice Mphahlele, Oluwatoyin Joseph Gbadeyan, and Viren Chunilall

7.1 INTRODUCTION

Because of the increasing population and high demand for environmentally friendliness, sustainable materials have sparked and motivated researchers to develop eco-friendly materials. Annually, roughly a hundred million natural fibers are harvested and consumed (1). These natural fibers offer advantages such as low density, biodegradability, inexpensive, abundance, and biocompatible as compared to non-biodegradable synthetic fibers such as metal, glass, and ceramic (2). Furthermore, synthetic fibers are problematic in the context of recycling while natural fibers are easy to recycle. To render sustainable and affordable yet eco-friendly materials, it has been revealed by many studies that the development of green composites using natural fibers and biodegradable polymers is pivotal. These green composites have shown superior mechanical properties rendering their services in various industrial uses such as packaging, toys, the energy sector, and biomedical (3).

Green composites are versatile materials with outstanding properties that are further being utilized in sectors such as sports, furniture, automobiles, shipping, and aviation (4). Green composites are developed to be environmentally friendly, addressing concerns about the ecological consequences of using green composite materials. They are either fully biodegradable or partially biodegradable (5). Fully biodegradable composites are obtained from a combination of biofibers and polymers derived from forestry feedstock or renewable agricultural feedstock. The production of complete biodegradable green composites-based starch and bamboo fiber showed outstanding properties and environmental friendliness (6). However, this composite suffers drawbacks such as moisture absorption and incompatibility between fibers and resin. To enhance the incompatibility of fibers and polymers, several treatment methods can be employed to enhance the compatibility between fiber and resin. The treatment of these fibers is influenced by the hydrophilic nature of the fibers which have poor interfacial bonding with hydrophobic polymer, leading to inadequate properties (7). Conventional treatment methods include physical, chemical, and biological treatment (8). Physical treatment methods alter the composite mechanical properties by

 DOI: 10.1201/9781003527275-7

altering the surface and structural characteristics of the fibers. Due to less effect on the chemical composition by physical treatment, greater interfacial adhesion between the polymer and fiber is observed (9). Physical treatments include, and not limited to, thermal, corona, and plasma. The degree of adhesion can also be increased by utilizing chemical treatments. Chemical treatments such as alkali treatment, peroxide treatment, acetylation, and bleaching can be utilized to increase the interfacial bonding between fibers and polymers (10). The most frequent chemical treatment method employed is alkali treatment, which is due to properties associated with low cost, ideal for effective surface treatment, and subsequently enhances the mechanical properties of fibers (10). Lastly, biological treatment, which involves the utilization of enzymes, can be used for the treatment of fibers. The enzymes mainly focus on the treatment of non-cellulosic parts of the fibers (11).

Green composites can be developed using conventional production processes such as compression molding, extrusion molding, hand lay-up, resin transfer molding, thermo-forming, pultrusion, and injection molding (12). The study focuses on the use of natural fibers for the production of green biocomposites. The use of various treatments and fabrication processes for the development of green composite was investigated. The developed green composite mechanical properties and applied applications were also comprehensively studied.

7.2 NATURAL FIBERS

Natural fibers are sourced from animals, plants, and minerals. They are hair-like structured materials utilized in composite development materials as reinforcements. Due to their satisfactory properties, they have been attracting enormous attention (13). Moreover, natural fibers play a vital role as a substitute for expensive and toxic synthetic fibers such as carbon, aramid, and glass. The classifications of natural fibers are depicted in Figure 7.1. Plant fibers are cellulosic fibers. They are classified into bast fibers, leaf fibers, fruit fibers, grass fibers, seed, and wood fibers (14). Bast fibers, which include flax, hemp, kenaf, and ramie, are extracted from the bast of the plant stem. Fibers collected from the leaves are called leaf fibers such as sisal, abaca, and banana. Fruit fibers include banana and coir, while grass fibers include bamboo and bagasse. Fibers such as kapok, rapeseed, and pine cone are considered seed fibers. Natural fibers such as softwood and hardwood are classified as wood fibers (15). Fibers extracted from animals include wool, silk, and fur. The naturally occurring fibers asbestos is considered a mineral fiber. Due to their hydrophilic properties which impair their mechanical properties in composites, it is paramount to functionalize or treat these fibers chemically or physically to improve their properties. These treatment methods are outlined in the following section.

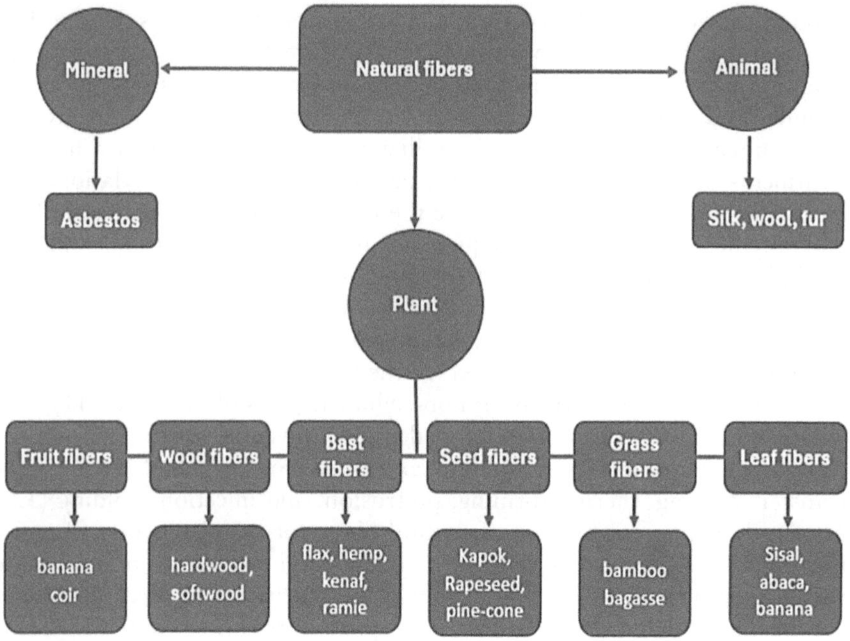

Figure 7.1 Classifications of natural fibers.

7.3 TREATMENT METHODS

The use of natural fibers in composite developments for various applications suffers drawbacks such as water absorption due to their high hydrophilic nature. Materials with moisture content decreased the mechanical properties of the composite (16). This is associated with the fact that fibers are also made up of non-cellulosic materials, including lignin, pectin, and hemicellulose, which have high moisture sensitivity, thus resulting in unfavorable mechanical properties. Furthermore, weather conditions such as rain, sun irradiation, and manufacturing processes may also affect the natural fibers mechanical properties. To curb the drawbacks of natural fibers, several treatment methods can be utilized to enhance the mechanical properties of natural fibers. Treatment methods such as physical and chemical processes are discussed.

7.3.1 Physical treatment

The surface of natural fibers can be treated to enhance the compatibility with polymers. The physical treatment process involves enhancing the interface between fiber and matrix without tempering the fiber's chemical composition (17). Several studies comprehensively investigated the effect of physical treatment on natural fiber-based composites (18). Physical treatment methods include plasma treatment, corona, UV treatment, and thermal treatment (17). Plasma treatment involves treating the cellulose fibers' surface at high voltage and low temperature while generating atmospheric pressure plasma (19). It is a very efficient process that does not alter the bulk properties of the fiber. Cold plasma therapy or corona therapy are two possible methods of treatment. Both approaches fall under the heading of plasma treatment when ionized gas is present and reacts with the current material surface through equivalent molecules with positive and negative charges. Frequency is utilized to differentiate between the two types of plasma during an electric discharge. The corona plasma is made by discharging an alternating current at ambient temperature, whereas high-frequency cold plasma was created using microwave energy. The modification of fiber surfaces depends on the kind of ionized gas and the duration of plasma exposure (20).

Another method of treating fibers to increase compatibility is corona treatment (20). By increasing surface energy and oxidizing the fiber surface, this method strengthens the link between the matrix that resists water and the water-loving fibers (15). The corona treatment system consists of a dielectric spacer made of quartz and two electrodes made of aluminum. During the corona treatment (Figure 7.2), the fiber surface experiences electric power through an air gap using the two electrodes. Only one side of the fibers will be sufficiently affected by the treatment, while the other side won't because each fiber must be placed on the electrode's surface; yet, composite materials enhance the fiber's entire lateral surface (16).

Ozone treatment process is also part of the physical treatment to enhance the properties of natural fibers. This treatment technique exposes the fibers on the surface to oxygen fluorine or an ozone atmosphere. The natural fibers are treated at different intervals in an ozone-rich, humid environment. The interaction of ozone with cellulose material is caused by the humid ozone state. The oxygen and ozone supply from the oxygen concentrator and

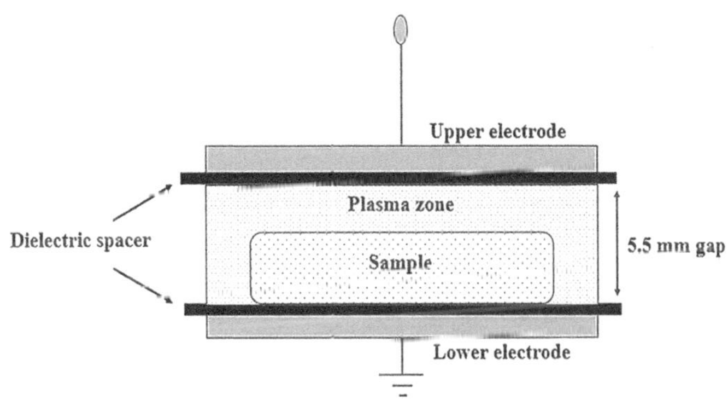

Figure 7.2 The process of corona treatment (21).

ozone generator, respectively, make up the humidification condition for the ozone treatment of fibers. The dry nature was exposed by ozone and eventually turned humid. Following treatment, the fibers were immersed in a non-ionic surfactant solution for an hour to eliminate any leftover ozone. Characterization of ozone-treated jute fibers reveals changes in their physical and chemical properties. Ozone treatment has also been depicted to cause lignin breakdown (19).

7.3.2 Chemical treatment

Chemical treatment on natural fibers enhances the compatibility between natural fibers and polymers via chemical reactions. The use of chemical treatment has gained enormous attention to evaluate and understand the effect of chemical treatment on the surface of the fibers. Furthermore, this is motivated by the hydrophilic nature of natural fibers and the hydrophobic nature of the polymers that suffer debonding due to unfavorable mechanical properties. The utilization of chemical treatment on the fibers decreases the moisture absorption behavior of fibers, thus improving adhesion with the matrix (22). Chemical treatment methods such as alkaline, malleated coupling agents, silane, acetylation, and permanganate treatment are widely used (23).

7.3.2.1 Alkaline treatment

Alkaline treatment is considered the most utilized treatment method due to its economic benefit, ease of use, and effectiveness in improving the interfacial adhesion between fiber and matrix. The method uses sodium hydroxide (NaOH) to treat the cellulosic structure of fiber (24). The fiber fragmentation speed and disaggregation increase due to alkaline treatment. During alkaline treatment, the sensitive OH groups are broken down and removed from the fibers thus generating a fiber cell-O-Na group with the reactive molecules as illustrated by equation (7.1). The removal of the OH group decreases the hydrophilic OH groups and a certain amount of pectin, lignin, oil, and hemicellulose within the fiber, which subsequently results in a fiber of high water resistance (25). Furthermore, a uniform and clean surface is obtained, which results in improved stress transfer.

$$\text{Fiber} - \text{OH} + \text{NaOH} = \text{Fiber} - \text{O} - \text{Na} + \text{H}_2\text{O} + \text{Impurities} \tag{7.1}$$

7.3.2.2 Malleated coupling agents

Malleated coupling agents can be utilized to provide significant interaction between matrix and fiber functional surfaces forming carbon-carbon bonding. Maleic anhydride reacts with the OH groups in the amorphous area of

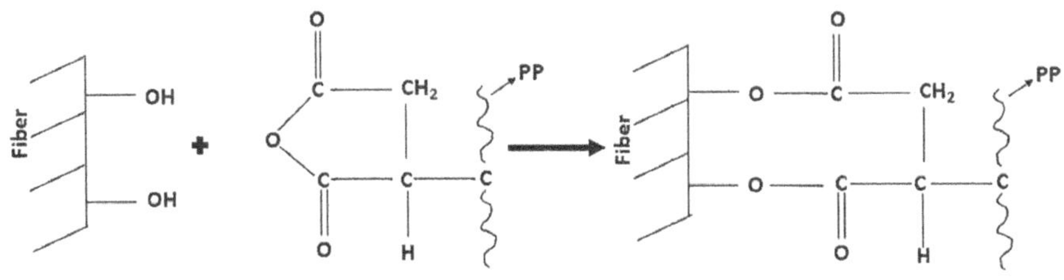

Figure 7.3 Bond formation reaction between natural fiber and anhydride ring of malleated polypropylene (MAPP).

the fiber cellulose structure to coat the fiber surfaces in long-chain polymers as shown in Figure 7.3. To reduce hydrophilic tendency, OH groups are excluded during coating on the surface of the fiber. The resulting surface leads to the formation of sufficient interlocking between fiber and polymer.

7.3.2.3 Silane treatment

Silane treatment involves coating micro-pores on the surface of natural fiber. A chemical link is formed via a siloxane bridge on the fiber surface and matrix during silane treatment. Initially, silanols are generated using hydrolysable alkoxy groups and moisture. The generated silanols on one end interact with the cellulose OH group while the matrix functional groups react with the other end as shown by equations (7.2) and (7.3). Consequently, a hydrocarbon chain is produced to prevent the fiber from expanding into the matrix and molecular continuity is created across the composite interface.

$$CH_2CHSi(OC_2H_5)_3 + H_2O \rightarrow CH_2CHSi(OH)_3 + 3C_2H_5OH \quad (7.2)$$

$$CH_2CHSi(OH)_3 + Fiber - OH \rightarrow CH_2CHSi(OH)_2 O - Fiber + H_2O \quad (7.3)$$

7.3.2.4 Acetylation treatment

The acetylation method involves modifying cellular structure by grafting acetyl groups using acid catalysts. The cellular structure of the fibers is modified by grafting acetyl groups using acid catalysts. Since acetic anhydride and acetic acid cannot separately react with fibers, the fibers are first soaked in acetic acid and then treated with acetic anhydride for three hours at a high temperature to speed up the reaction. Under this method, the OH and carboxylic/anhydride groups of the natural fibers undergo an esterification process (26). The chemical reaction between the fiber and acetic anhydride is illustrated by equation (7.4).

$$\text{Fiber} - \text{OH} + \text{CH}_3 - \text{C}(=\text{O}) - \text{O} - \text{C}(=\text{O}) - \text{CH}_3$$
$$\rightarrow \text{Fiber} - \text{OCOCH}_3 + \text{CH}_3\text{COOH} \quad (7.4)$$

7.3.2.5 Permanganate treatment

Permanganate treatment involves the reaction of potassium permanganate ($KMnO_4$) with the natural fiber OH group and lignin components of the fibers to improve adhesion between the fibers and the matrix, enhancing the fiber's thermal stability. Utilizing permanganate to natural fibers enhances their chemical interlocking at the matrix-natural fiber interface (27). The chemically treated jute fibers were investigated. The fibers were first suspended in NaOH solution to increase compatibility between fiber and matrix, and subsequently immersed in permanganate acetone solution. A significant weight reduction was observed due to the removal of lignin, oil, wax, hemicellulose, and pectin from the fibers. The resulting mechanical properties revealed improved tensile strength by 20% (27).

7.4 GREEN COMPOSITE MANUFACTURING TECHNIQUES

Green composites can be developed using conventional production processes, which include, and not limited to, compression molding, injection molding, extrusion molding, hand lay-up, and resin transfer molding (12).

7.4.1 Compression molding

The production of high-performance and lightweight-based thermoplastic and natural fiber composites has been achieved using the compression molding process shown in Figure 7.4. The process involves placing your pre-heated materials in the fabrication molding cavity. The core side of the mold then compresses and deforms the materials while applying tremendous pressure to the cavity. The high pressure is kept until the composite solidifies before opening the mold and extracting the composite. Several parameters, such as the amount of feedstock, applied pressure, and heating and cooling time, need to be monitored for the effective fabrication of the composites (28).

7.4.2 Injection molding

The most popular method for producing composites in large quantities is injection molding, as shown in Figure 7.5. The first step is to place the polymers in the hopper in the form of pellets or granules and heat them in the

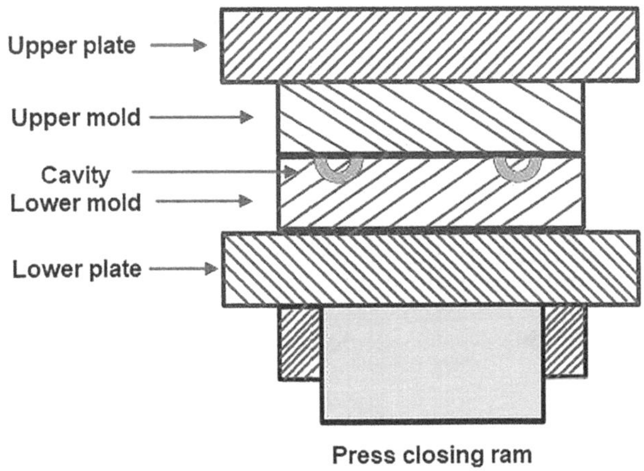

Figure 7.4 Compression fabrication process.

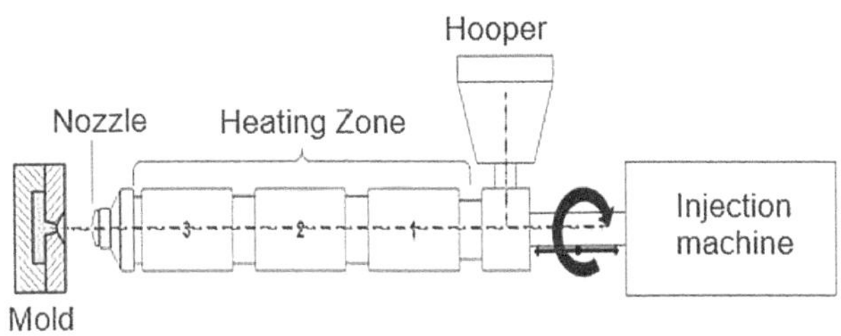

Figure 7.5 The principle of injection fabrication process.

heating zone until they melt. After that, the molten components are maintained in the mold and injected into a chamber created by a split die mold. After cooling down, the melt solidifies and the mold is opened (12).

7.4.3 Extrusion molding

One of the most popular techniques in the production of natural fiber composites is extrusion molding, as shown in Figure 7.6. The great stiffness and strength of the resulting composites, along with the ease of composite manufacturing, make this process the favored one. The first step in this process is to store the thermoplastic material in a hopper in the form of pellets or granules. After that, they are placed in a heated barrel in the heating zone section to melt. The resulting material was then hardened through water bath. The last step includes grinding the material with a mechanical grinder (29).

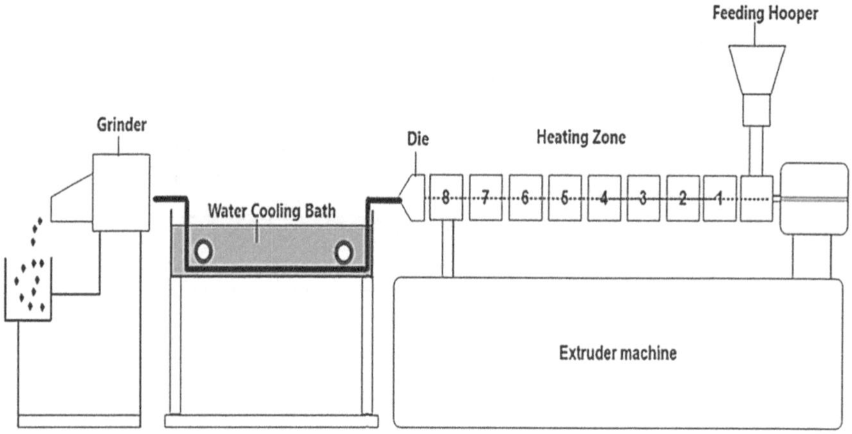

Figure 7.6 The representation of the extrusion process of thermoplastic material (29).

7.4.4 Hand lay-up

Hand lay-up is the most popular, straightforward, and affordable method for producing green composites. Using this method, which is depicted in Figure 7.7, the open mold is first treated with a release agent that acts as an anti-adhesive agent before the fibers are inserted into the mold. Resins are poured over the fibers and then brushed on with either a roller or brush. Layers are built on top of layers to get the appropriate thickness in lay-up.

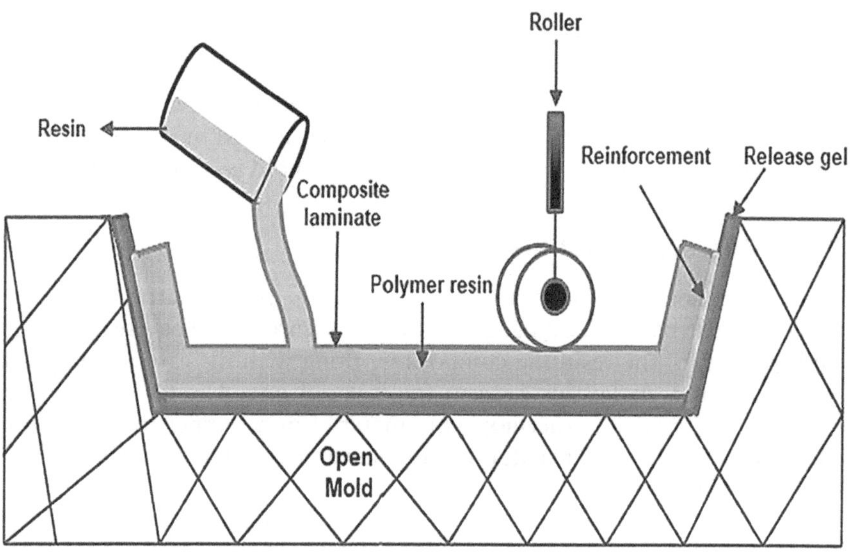

Figure 7.7 Illustration of hand lay-up (30).

The trapped air in the laminate is manually eliminated using rollers or squeegees (14).

7.4.5 Resin transfer molding (RTM)

Rather than feeding the polymer into an open mold, resin transfer molding (RTM) loads the resin into a holding chamber after it has been pre-heated. Large component production at medium volume is best suited for this technology. After the textile layers are positioned beforehand in the solid mold, the preforms are impregnated with resin through injection. Vacuum is frequently utilized to reposition the resin into the cavity and prevent air bubbles (31). The green composites materials fabricated through various manufacturing techniques are then investigated through various testing methods to evaluate their mechanical properties.

7.5 MECHANICAL PROPERTIES

Green composite materials have various mechanical properties such as impact strength, tensile, flexural, and compressive properties. Different mechanical properties are exhibited by different green composites.

7.5.1 Tensile properties

The interfacial bonding between the resin and fibers mostly influences the tensile properties of natural fiber composites. Composites have better tensile properties when the fiber and resin are modified chemically and physically (32). The fiber volume percentage in the matrix resin has a significant impact on the tensile properties of natural fiber composites. It is generally true that when the fiber volume fraction increases below an optimal value, the load is distributed over a greater number of fibers. In this case the matrix can support the applied load after the fiber fracture, and the composite may have a greater tensile strength as a result (32). Contrary to fiber volume percentage, brittle fracture may develop in the fibers with additional increases in fiber volume fraction beyond the optimal level, and the matrix is unable to sustain the additional load from the fibers (32). The poor tensile strength eventually causes the entire composite to break under these circumstances. Numerous factors, including fiber degradation, incorrect manufacturing process, and incompatibility between the fibers and matrix, might contribute to less favorable tensile properties (5).

7.5.2 Flexural properties

Flexural stiffness is one of the most important factors in determining how resistant a composite is to bending deformation. The composite material's modulus is one of the factors that affect flexural properties (16). The flexural strength of green composite is increased by increasing the fiber content to the ideal level. Further addition of natural fiber above the ideal level results in less favorable flexural strength because of flaws in the fiber's wetting that might cause stress concentration areas in the composite materials (16). The composite flexural modulus likewise rises with an increase in fiber content. Green composite resistance to crack propagation, or fracture toughness, is primarily determined by the matrix's intrinsic qualities, the fiber volume percentage, and the interfacial bonding strength (33).

7.5.3 Compressive properties

The fiber volume fraction and reinforcing design of natural fiber composites have a significant impact on their compressive behavior (34). Due to fewer voids in the composite when the fiber volume percent is increased below the ideal level, the composite exhibits higher compressive strength. Another factor contributing to increased compressive strength is strong compatibility and good uniformity between the fibers and polymers (35). The compressive strength decreases when the fiber volume fraction rises above the ideal level. In general, with every 0.5% increase in fiber volume, compressive strength decreases by about 8.5% when the fiber volume fraction is higher than 1.5% (35). Nonetheless, research has demonstrated that when the fiber volume fraction increased, the compressive strength of natural fiber composites progressively decreased.

7.5.4 Impact strength

The degree of adhesion between the fiber and matrix determines how natural fiber composites behave under impact. Throughout the course of the natural fiber composites' service life, this characteristic is crucial for the sustainability of the composite. The impact strength measures the resistance of the material to initiate cracks that can propagate throughout the material. Through modification techniques, natural fiber composites can have their impact characteristics improved. Fiber pull-out, better bonding, energy absorption, and degree of adhesion are a few factors that affect a composite's impact strength (32).

7.6 MODELING AND SIMULATION

The mechanical behavior of green composites can be predicted using computational models, which are the shear lag model, the Halpin–Tsai equation, the Hirsch model, the laws of mixture, series and parallel, and the Bowyer–Bader model (36). Elastic modulus of the fiber (E_f), elastic modulus of the matrix (E_m), tensile strength of the fiber (σ_f), tensile strength of the matrix (σ_m), fiber volume fraction (V_f), and matrix volume fraction (V_m) are the basic parameters utilized in these models. The green composite interface and micro-defects in the polymer are not taken into account in these models (37).

7.7 GREEN COMPOSITE APPLICATIONS

Green composites are utilized in various applications owing to their advantageous properties such as durability, lightweight, and integrity. These properties increased its applications in industries such as aerospace, automotive, medical, electrical field, sports, construction, and packaging. Generally, green composites are applied in all advanced engineering structures. Currently, several industries encourage the use of green composites because of environmental and economic benefits (38). This section focuses on various green composite applications.

7.7.1 Aerospace applications

The percentage of natural composites in the airspace is about 50%. The Wright Brothers' Flyer 1 was the first aircraft to employ natural fiber components, including wood and textiles. Because natural composites have the ability to lower an aircraft's weight, they are now utilized in the development of helicopters, launch vehicles, civil transport aircraft, cockpits, small and large, military fighter aircraft, satellites, and missiles. Natural composites are used in the fabrication of the aircraft's rudder, airbrakes, engine cowlings, turbine engine fan blades, spoilers, keel beam, LG doors, elevators, propellers, wing ribs, rear bulkhead, main wings, and interior components (39).

7.7.2 Automotives applications

Green composite materials have been employed in recent decades to create lighter, safer, easier-to-recycle, and more fuel-efficient automobiles. Natural fibers are used to create the car's principal parts, which include the exterior panels, dashboards, interior door panels, wheels, seat backboards, trunks, roofs, mats, and hatches (40). The yearly need for natural fibers in the automotive sector is rising daily; in 2005, the sector utilized over 15,000 tons of natural fibers (41).

Natural fibers are used by German automakers such as BMW, Volkswagen, Mercedes, Audi Group, Opel, Ford, and Daimler Chrysler to fabricate door panels, boot linings, headliner panels, seat backs, molded feet, and noise insulation panels. Furthermore, natural fibers are used to make hat racks, seat backs, boot linings, side and rear door panels, and spare tire linings. Ford uses natural fiber for door panels and boot lining. Mercedes-Benz is made of natural fibers for the internal engine cover, sun visor, engine insulation, interior insulation, wheel box, bumper, and roof cover (41). Other automotive companies, such as the national carmaker of Malaysia, Proton Company, and the American auto industry, Cambridge Industry, manufacture various car components using natural composites, primarily polypropylene or polyester, and fibers such as hemp, jute, flax, or sisal. Global demand for natural fiber-reinforced composites in the automotive industry is rising quickly due to marketing demands more than technological ones (39).

7.7.3 Medical applications

Currently, composite materials have been utilized extensively in medical applications. Medical devices are made of green composite materials, which interact with the biological system to enhance its quality. The green composite material has been employed in numerous ways over the last few decades. It has occasionally been used to develop surgical techniques and sterilization processes. In orthopedic applications, green composite materials have been discovered to be useful, specifically in bone grafting, bone fixation plates, hip joint replacement, and bone cement (42).

In clinical dentistry, composite materials are also utilized for anterior and posterior tooth restoration. The tooth that is broken or missing is replaced using green composites. These green composites can further be utilized to enhance the quality of life of patients. Examples of these materials include intraocular lenses, vascular grafts, pacemakers, artificial hearts, biosensors, and heart valves. They are also occasionally used to restore bodily functions, such as those of tissues or organs (39).

7.7.4 Electrical field applications

An essential requirement for the electric field is that green composite materials should have a low dielectric constant, high strength, high thermal conductivity, low thermal expansion, and high/low electrical conductivity. Composite materials made of natural fibers that function in an electric field are not only utilized as insulators but also as conductors. Occasionally, the composite materials are used as switches, home plugs, terminal circuits, boards, and connectors (39). The electrical characteristics of pineapple-reinforced polyethylene composite were investigated. The composite's electrical characteristics demonstrated its electrically anisotropic nature (39).

7.7.5 Sports applications

Due to their many advantages over other materials, including weight, shaping, ease of maintenance, superior thermal stability, high strength, resistance to abrasion, resistance to friction, low vibration attenuation, and high design freedom, green composite materials are now commonly used in sports equipment. Nowadays, a lot of sports equipment is constructed with green composite materials. For example, natural fiber composites are used to make wind-surfing, fishing rods, golf club heads, golf clubs, swords, skis, tennis boards, climbing ropes, slats, and other equipment (43). With the use of green composites in sports equipment, the mechanical properties increased by about 30% to 50%. In comparison to other sports equipment, it is lightweight and has good mechanical performance (43).

7.7.6 Building and construction applications

Green composites are utilized in construction materials. A significant role for green composite materials is played in the building and construction industry. Numerous construction industries have used green composite materials, including high-rise buildings, industrial supports, long-span roof structures, tanks, lightweight doors, furniture, windows, lightweight buildings, and entire bridge systems (44). Coir fibers can be utilized to manufacture wall panels, partitions, ceilings, and roofing boards when combined with polymers to create an even better composite. Moreover, rice husk-based composite can be used to produce wall panels, flooring, partition walls, roofing, and door panels (39).

7.7.7 Packaging applications

The majority of resins used in packaging materials are highly expensive; however, adding natural fibers to green resins not only lowers the material's cost but also enhances its strength, stiffness, and ability to withstand heat without compromising its biodegradability (39). Packaging materials made of natural fibers need to be environmentally friendly.

7.8 CONCLUSIONS

Natural fibers are a renewable, biodegradable, and environmentally beneficial raw material. Both mechanical and thermal insulation properties are benefits of natural fibers. Green composites, which are fabricated using natural fibers and bio-based polymers, are investigated. This chapter provides an overview of green composites. The appropriate fabrication methods and different treatments to improve the performance for long-term durability are crucial to the success of these materials. However, enhanced mechanical properties will be achieved. Also, computation modeling pre-determines a combination of materials subjected to simulation testing for fabricating working composites with improved properties for different applications. In summary, due to their excellent properties, green composites are being used in applications that include and are not limited to automotive, aero space, building and construction, medical, and packaging. Even though significant properties are achieved using green composites, further research is still needed to overcome the drawbacks of these green composites. Overall, green composites serve as a substitute for synthetic composites mainly because of their abundance, recyclability, biodegradability, and environmental friendliness.

REFERENCES

1. Townsend T, Sette J 2016. Natural fibres and the world economy. *Natural Fibres. Advances in Science and Technology Towards Industrial Applications.* Springer, 381–91.
2. Khalid YM, Rashid, AA, Zia UA, Waqas A, Hassan A, Asad AZ 2021.Engineering natural fiber reinforced composites. *Sustainable Materials for Emerging Applications*, 63–75. Doi:10.1016/j.rineng.2021.100263
3. Lima DC, de Melo RR, Pimenta AS, Pedrosa TD, de Souza MJC, de Souza EC 2020. Physical–mechanical properties of wood panel composites produced with Qualea sp. sawdust and recycled polypropylene. *Environmtal Science Pollution Research* 27, 58–65.
4. Amjad F, Abbas W, Zia-Ur-Rehman M, Baig SA, Hashim M, Rehman HU 2021. Effect of green human resource management practices on organizational sustainability: The mediating role of environmental and employee performance. *Environmental Science Pollution and Research* 28, 191–206.
5. Rahman MA, Haque S, Manokar M, Mohamed A, Kamaludeen B 2023. A review of environmental friendly green composites: Production methods, current progresses, and challenges. *Environmental Science and Pollution Research* 30, 5–29. Doi:10.1007/s11356-022-24879-5
6. Nguegang W, Zhang J, Semple KE, Chen M, Xia Y, Dai C 2023. Bamboo-based composites: A review on fundamentals and processes of bamboo bonding. *Composite Part B* 235, 76–96. Doi:10.1016/j.compositesb.2022.109776
7. Parikh HH 2023. Tribology of plant-based natural fiber reinforced polymer matrix composites: A short review. *Journal of Natural Fibers* 20, 2172639–55. Doi:10.1080/15440478.2023.2172639
8. Khoaele KK, Gbadeyan OJ, Chunilall V, Sithole B 2023. The devastation of waste plastic on the environment and remediation processes: A critical review. *Sustainability* 15, 5233–47.
9. Nurazzil, N, Harussani M, Aisyah H, Ilyas R, Norrrahim M, Khalina A, 2021. Treatments of natural fiber as reinforcement in polymer composites: A short review. *Functional Composites and Structure* 3, 024002–18.
10. Cai M, Takagi H, Nakagaito AN, Li Y, Waterhouse GIN 2016. Effect of alkali treatment on interfacial bonding in abaca fiber-reinforced composites. *Composites Part A: Applied Science and Manufucturing* 90, 589–97. Doi:10.1016/j.compositesa.2016.08.025
11. Birniwa AH, Sa S, Ali M, Edrees R, Mohammad A, Jagaba AH, et al. 2023. Recent trends in treatment and fabrication of plant-based fiber-reinforced epoxy composite: A review. *Composite Science* 7, 120–53.
12. Mann GS, Singh LP, Kumar P, Singh S 2020. Green composites: A review of processing technologies and recent applications. *Journal of Thermoplastic Composite Materials* 33, 1145–71.

13. Akil HM, Omar MF, Mazuki AAM, Safiee S, Ishak ZAM, Abu Bakar A 2011. Kenaf fiber reinforced composites: A review. *Materials and Design* 32, 4107–21. Doi:10.1016/j.matdes.2011.04.008
14. Gholampour A, Ozbakkaloglu T 2020. A review of natural fiber composites: Properties, modification and processing techniques, characterization, applications *Journal of Materials Science* 55, 829–92 Doi:10.1007/s10853-019-03990-y
15. Ramesh M, Palanikumar K, Reddy KH 2017. Plant fibre based bio-composites: Sustainable and renewable green materials. *Renewable and Sustainable Energy Reviews* 79, 558–84. Doi:10.1016/j.rser.2017.05.094
16. Faruk O, Bledzki AK, Fink HP, Sain M 2012. Biocomposites reinforced with natural fibers: 2000–2010. *Progress in Polymer Science.* 37, 1552–96. Doi:10.1016/j.progpolymsci.2012.04.003
17. Nurazzi NM, Harussani MM, Aisyah HA, Ilyas RA, Norrrahim MNF, Khalina A, 2021. Treatments of natural fiber as reinforcement in polymer composites: A short review. *Functional Composites Structructures* 3, 024002–18.
18. Ahmad R, Hamid R, Osman SA 2019. Physical and chemical modifications of plant fibres for reinforcement in cementitious composites. *Advances in Civil Engineering* 1, 6–25.
19. Marais S, Gouanvé F, Bonnesoeur A, Grenet J, Poncin-Epaillard F, Morvan C 2005. Unsaturated polyester composites reinforced with flax fibers: Effect of cold plasma and autoclave treatments on mechanical and permeation properties. *Composites Part A: Applied Science and Manufacturing* 36, 975–86.
20. El-Naggar HI, Al-Halim MAA, Gaballa AS, Hassouba MA 2021. Surface improvement of cotton/polyester blend textile using DC air glow discharge plasma. *Physica Scripta* 96(12), 125712–25.
21. Ragoubi M, Bienaimé D, Molina S, George B, Merlin A 2010. Impact of corona treated hemp fibres onto mechanical properties of polypropylene composites made thereof. *Industrial Crops and Products* 31, 344–9.
22. Faruk O, Bledzki AK, Fink HP, Sain M 2014. Progress report on natural fiber reinforced composites. *Macromolecular Materials Engineering* 299, 9–26.
23. Jagadeesh P, Puttegowda M, Mavinkere Rangappa S, Siengchin S 2021. A review on extraction, chemical treatment, characterization of natural fibers and its composites for potential applications. *Polymer Composites* 42, 6239–64.
24. Samaei SE, Mahabadi HA, Mousavi SM, Khavanin A, Faridan M, Taban E. The influence of alkaline treatment on acoustical, morphological, tensile and thermal properties of Kenaf natural fibers. *Journal of Industrial Textiles* 51, 8601S–25S.
25. Iannace S, Ali R, Nicolais L 2001. Effect of processing conditions on dimensions of sisal fibers in thermoplastic biodegradable composites. *Journal of Applird Polymer Science* 79, 1084–91.
26. Zaman HU, Khan RA 2021. Acetylation used for natural fiber/polymer composites. *Journal of Thermoplastic Composite Materials* 34, 3–23.
27. Khoaele KK, Gbadeyan OJ, Chunilalla V, Sithole B 2023. A review on waste wood reinforced polymer composites and their processing for construction materials. *International Journal of Sustainable Engineering* 16, 1–13. Doi:10.1080/19397038.2023.2214162

28. Araújo JR, Waldman WR, De Paoli MA 2008. Thermal properties of high density polyethylene composites with natural fibres: Coupling agent effect. *Polymer Degradradation and Stability* 93, 1770–5.
29. Hietala M 2013. Extrusion processing of wood-based biocomposites. (Doctoral dissertation, Luleå tekniska universitet), 1–116.
30. Ed M, Razzaq A, Erfan S, Sanaul M 2021. Mechanical properties of biofiber/glass reinforced hybrid composites produced by hand lay-up method: A review. *Materials Engineering Research* 3, 144–55.
31. Sèbe G, Cetin NS, Hill CAS, Hughes M. RTM2000. Hemp fibre-reinforced polyester composites. *Applied Composite Material* 7, 341–9.
32. Ku H, Wang H, Pattarachaiyakoop N, Trada M 2011. A review on the tensile properties of natural fiber reinforced polymer composites. *Composite Part B* 42, 856–73.
33. Kinloch AJ, Taylor AC, Techapaitoon M, Teo WS, Sprenger S 2015. Tough, natural-fibre composites based upon epoxy matrices. *Journal of Material Science* 50, 6947–60.
34. Węcławski BT, Fan M, Hui D 2014. Compressive behaviour of natural fibre composite. *Composite Part B: Engineering* 67, 183–91.
35. Afroughsabet V, Ozbakkaloglu T 2015. Mechanical and durability properties of high-strength concrete containing steel and polypropylene fibers. *Construction and Building Materials* 94, 73–82. Doi:10.1016/j.conbuildmat.2015.06.051
36. Mulenga TK, Ude AU 2021.Techniques for modelling and optimizing the mechanical properties of natural fiber composites: A review. *Fibers* 9, 6–23.
37. Ilczyszyn F, Cherouat A, Montay G 2014. Effect of hemp fibre morphology on the mechanical properties of vegetal fibre composite material. *Advanced Material Research* 877, 485–9.
38. Sun Z 2018. Progress in the research and applications of natural fiber-reinforced polymer matrix composites. *Science* and *Engineering* of *Composite* 25, 835–46.
39. Keya KN, Kona NA, Koly FA, Maraz KM, Islam MN, Khan RA 2019. Natural fiber reinforced polymer composites: History, types, advantages, and applications. *Material Engineering Research*1(2), 69–87.
40. Koronis G, Silva A, Fontul M 2013. A review of adequate materials for automotive applications. *Composites: Part B Green composites* 44, 120–7. Doi:10.1016/j.compositesb.2012.07.004
41. Pickering KL, Efendy MGA, Le TM 2016. A review of recent developments in natural fibre composites and their mechanical performance. *Composites Part A* 83, 98–112.
42. Naik N, Shivamurthy B, Thimmappa BHS, Jaladi G, Samanth K, Shetty N 2023. Recent advances in green composites and their applications. *Engineering Sciences* 21, 779–95.
43. Jinliang W 2012. Application of composite materials on sports equipments. *Applied Mechanics and Materials* 156, 903–6.
44. Lesko JJ, Machida A, Rizkalla SH, Asce F, Triantafillou TC, Asce M. Fiber-reinforced polymer composites for construction: State-of-the-art review. *American Society of Civil Engineering* 6,73–87.

Chapter 8

ANN applications in polymer tribology

A concise review

Gaurang Patel, Vijaykumar Chaudhary, and Kawaljit Randhawa

8.1 INTRODUCTION

Engineering challenges become more and more complex and difficult to solve. Conventional statistical methods have limitations in attending complex engineering problems in nature [1]. Many researchers in the past have tried to visualize and solve the machine data through conventional statistical tools, but problems occur when fitting of the data is nonlinear [2]. Linear regression analysis can solve problems that are linear in nature but cannot give accurate results in case of nonlinear relations of input parameters with outputs. Tribological properties are more complexly influenced by experimental conditions such as pressure, load, speed, sliding distance, temperature, time, orientation of fibers, surface topology, relative surfaces, nature of contacting surface, environmental operating conditions, etc. [3]. Though many researchers attempt to develop mathematical models through conventional models, they cannot achieve accuracy. However, it can be achieved by tuning machine learning models for tribological applications [4–7].

A variety of machine learning models are available in the market, each with its own computing ability and adaptability to different data types. ML models try to fit with given datasets by understanding the nature of the data and the distribution of the same called training [8]. These ML models have the tendency to pass through each point of the dataset, provided sometimes they may deflect the predictions of outputs in the wrong direction. For avoiding this type of problem, data pre-processing is much important before deploying ML models. Generally, ML methods can be divided into supervised learning, unsupervised learning, and reinforcement learning [8], as shown in Figure 8.1. The common applications of ML algorithms in different areas are shown in Figure 8.2.

DOI: 10.1201/9781003527275-8

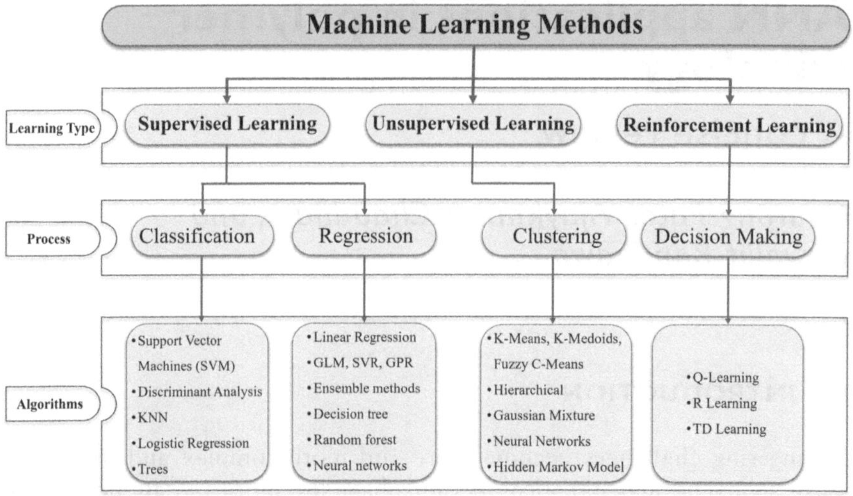

Figure 8.1 Machine learning methods.

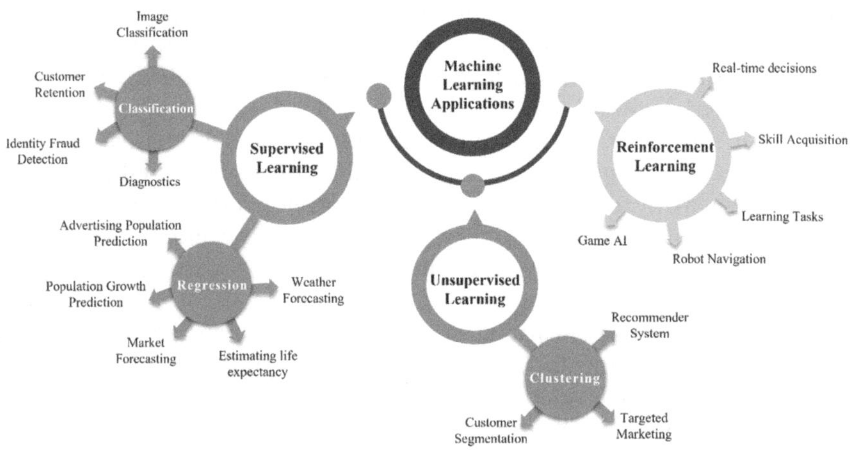

Figure 8.2 Applications of machine learning methods.

8.2 ARTIFICIAL NEURAL NETWORK

The architecture of the ANN is based on nodes called artificial neurons, which act the same as the neurons of the biological brain [9]. The biological brain is much more complex and quickly reoriented toward improving empirical results. One neuron in the brain is connected to others in various ways, so the outputs of some neurons become inputs for other neurons. As shown in Figure 8.3, the dendrites collect the inputs and pass through the axon, and finally the synaptic terminals release the outputs. ANN is a

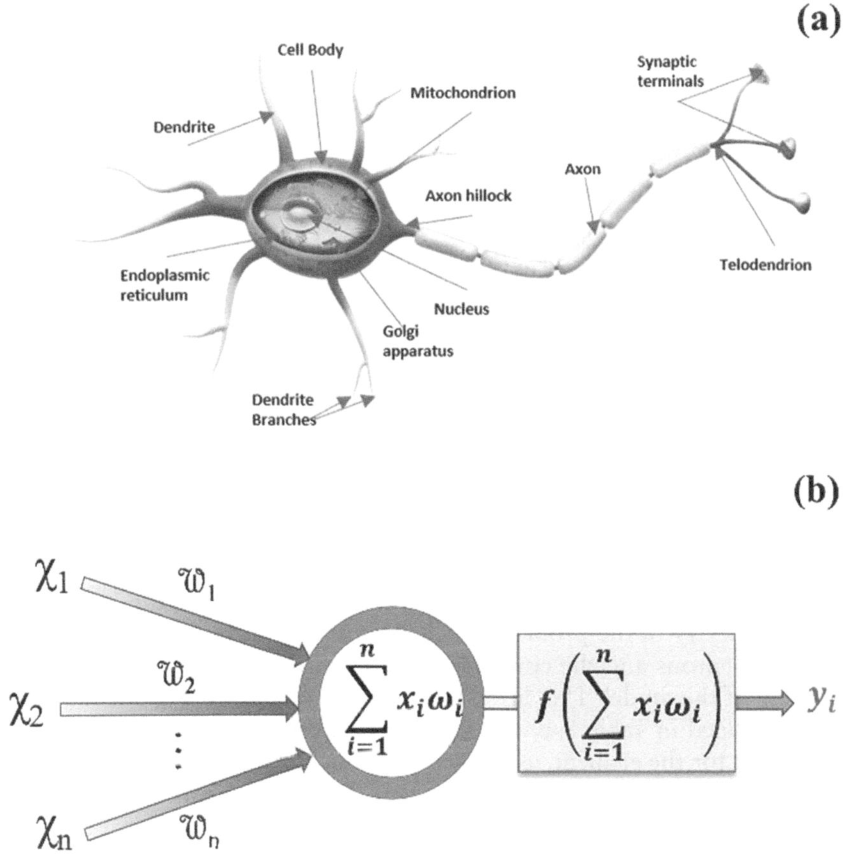

Figure 8.3 (a) Biological brain neuron, (b) artificial neuron.

widely accepted model among research communities dealing with complex problems [10–16]. The interpretation of the working of the ANN model is in terms of interrelations between the input parameters and output parameters, quite difficult and so widely accepted, likely to be a black-box model [17]. Figure 8.4 demonstrates the general construction of an ANN, accounting for the major affecting parameters as inputs, with the output parameters being the coefficient of friction (COF) and specific wear rate (SWR).

8.2.1 Backpropagation of ANN

The backpropagation algorithm is used for function approximation, for pattern association, and classification of multilayer feed-forward networks, such as ANN [7, 16, 18–20]. The computation of derivatives of error at each forward propagation with respect to weights and biases is done by the backpropagation algorithm. The ANN model can be trained by adjusting weights

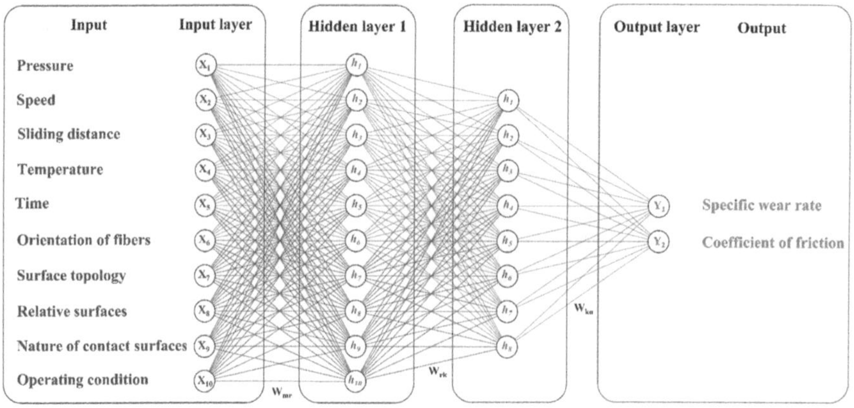

Figure 8.4 Artificial neural network for tribology.

followed by backpropagation and calculation of the network error associated, and by forward propagation of the input training pattern. Generally, the architecture of an ANN develops with three layers in sequence of the input layer, output layer, and hidden layer [8]. According to the application and complexity of the problem, the hidden layer can be added. Each layer contains neurons and the connections between these generate the network function of the model. The different activation functions used for the ANN are mentioned in Table 8.1. The relation between the input parameter and the output for the element, as shown in Figure 8.3(b), can be described as:

$$y_i = f\sum_{i=1}^{n} x_i w_i \tag{8.1}$$

The learning of the model is based on a gradient search method and the accuracy is measured by the least sum of squared errors.

$$E = \sum_{p=1}^{p} \left(t_p - a_p\right)^2 \tag{8.2}$$

Here t_p and a_p are the predicted (target) output and measured (actual) output, respectively, and E is the average of all patterns in the training set. The learning algorithm used for adjusting weights up to the desired minimum level of error achieved, in the case of the minimization process, is shown in equation (8.3), and the correcting weight equation is shown in equation (8.4).

$$w_i(t+1) = w_i(t) + \Delta w_i(t) \tag{8.3}$$

Table 8.1 Activation functions used in ANN

Formula	*Range*	*Name*
$f(x)=\max(0,x)$	$(0,\infty)$	ReLU
$f(x)=\begin{cases}0.01x, & x<0\\ 0, & x\geq 0\end{cases}$	$(-\infty,\infty)$	Leaky ReLU
$f(x)=\sin(x)$	$(-1,1)$	Sinusoidal
$f(x)=\frac{1}{1+e^{-x}}$	$(0,1)$	Logistic
$f(x)=e^{-x}$	$(0,\infty)$	Exponential
$f_i(\vec{x})=\frac{e^{x_i}}{\sum_{j=1}^{j}e^{x_j}}$	$(0,1)$	Softmax
$f(x)=e^{-x^2}$	$(0,1)$	Gaussian
$f(x)=x$	$(-\infty,\infty)$	Linear

$$\Delta w_i(t+1)=-\eta\frac{\partial E}{\partial w_i}+\mu\Delta w_i(t) \tag{8.4}$$

where $w_i(t)$ and $\Delta w_i(t)$ are the weights allotted to the correct answer and the correction of the weight at t^{th} iteration step. The learning rate η is the adjustment of correction each time, and the momentum factor μ aids in rapid convergence by reducing momentum.

8.2.2 Overfitting problem

Deep neural networks are designed well enough to learn any data from different domains. ANNs have the freedom to fit with different types of datasets, making it more powerful, but as the model tries to fit with many details, it also trains over noise from the training data called overfitting [21]. In overfitting, the model fails to identify the correct patterns present in the training dataset, and so the final model performance is not accurate. The possible reasons for overfitting of ANNs [22] are first the training dataset size is small: when the ANN tries to train with a small dataset, it will try to pass with all points exactly and memorize each point and so fail to understand the general trend of the training dataset. Second, the predictions on

Noisy Data: the complex problems having too many parameters may create a dataset very noisy. Once the model tries to fit, this type of data becomes overfitted and does not reflect correct outputs [22].

8.3 ARTIFICIAL NEURAL NETWORKS FOR POLYMER TRIBOLOGY

Nikhil Sharma et al. [23] have experimented on neat glass-fiber reinforced epoxy composite with graphene nanoplatelets of 0.5 and 1 by wt.% and determined the SWR and COF. The ANN model was developed by modifying the count of neurons and was finally set to 13 for the prediction of the output parameters shows the best results as shown in Figure 8.5 with MSE of 0.010315 for SWR and 0.000119 for COF.

Researchers [24] have studied tribological properties for medical grade UHMW-PE polymers under dry sliding and Hank's balanced salt solution (HBSS) conditions. The ANN model was developed using nine hidden, two input, and two output layers and deployed to predict the values of wear and friction. For the learning process of the model, η and μ were set default to 0.7, and 300,000 iterations were used. The formulations of error measuring parameters used in the study are as follows:

$$\text{MRE} = \frac{1}{n}\sum_{i=1}^{n} abs(X) \tag{8.5}$$

$$\text{STD} = \sqrt{\frac{\sum_{i=1}^{n}\left(X - \bar{X}\right)^2}{n-1}} \tag{8.6}$$

where X is $\left(O_p - O_e\right)/O_e$. O_e and O_p are the output parameters by experimentation and predicted by ANN, respectively.

$$R^2 = 1 - \left[\frac{\sum_{i=1}^{n}\left(x_i - y_i\right)^2}{\sum_{i=1}^{n}\left(y_i\right)^2}\right] \tag{8.7}$$

where the predicted output is y_i and the experimental output is x_i.

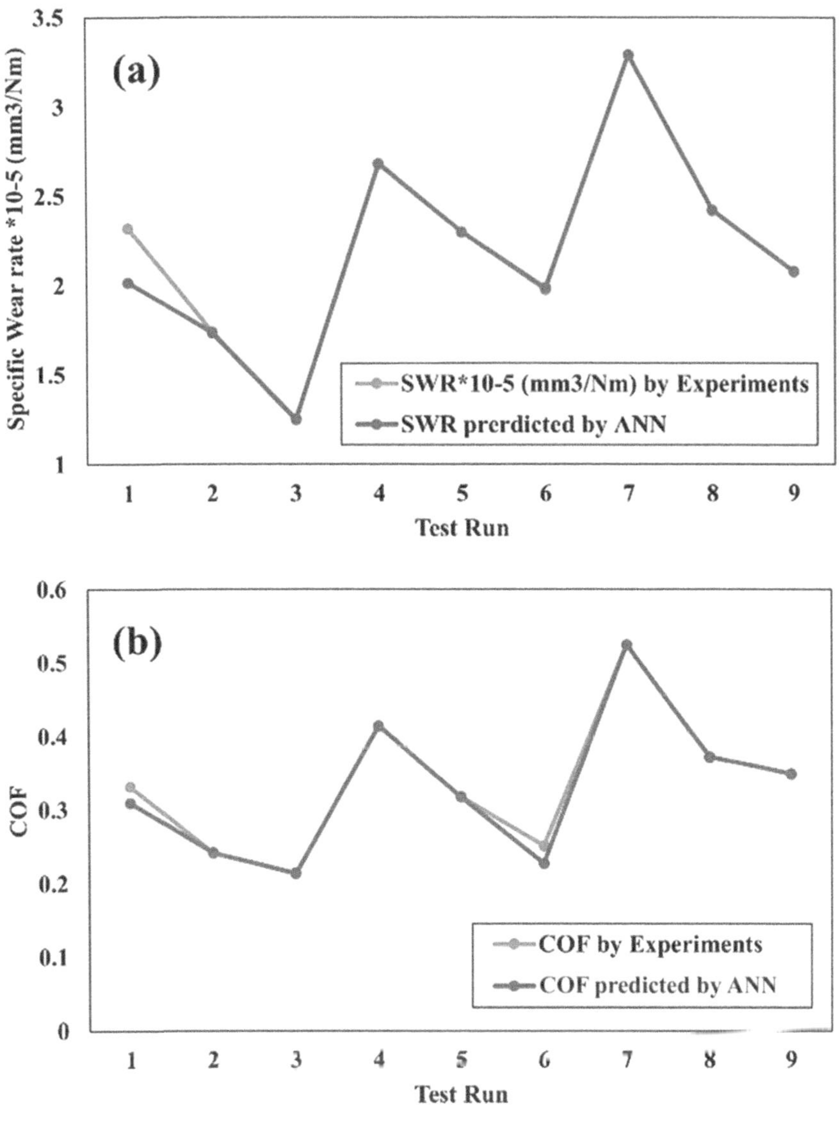

Figure 8.5 Comparison of the output values by experiments and by ANN (a) SWR (b) COF

The UHMW-PE performed well under HBSS lubrication conditions and showed minimum COF and SWR as shown in Figure 8.6(a) and (b). The developed ANN model predicted the output of COF with 0.194578% of MRE and SWR with 0.0042265% of MRE, as shown in Figure 8.7.

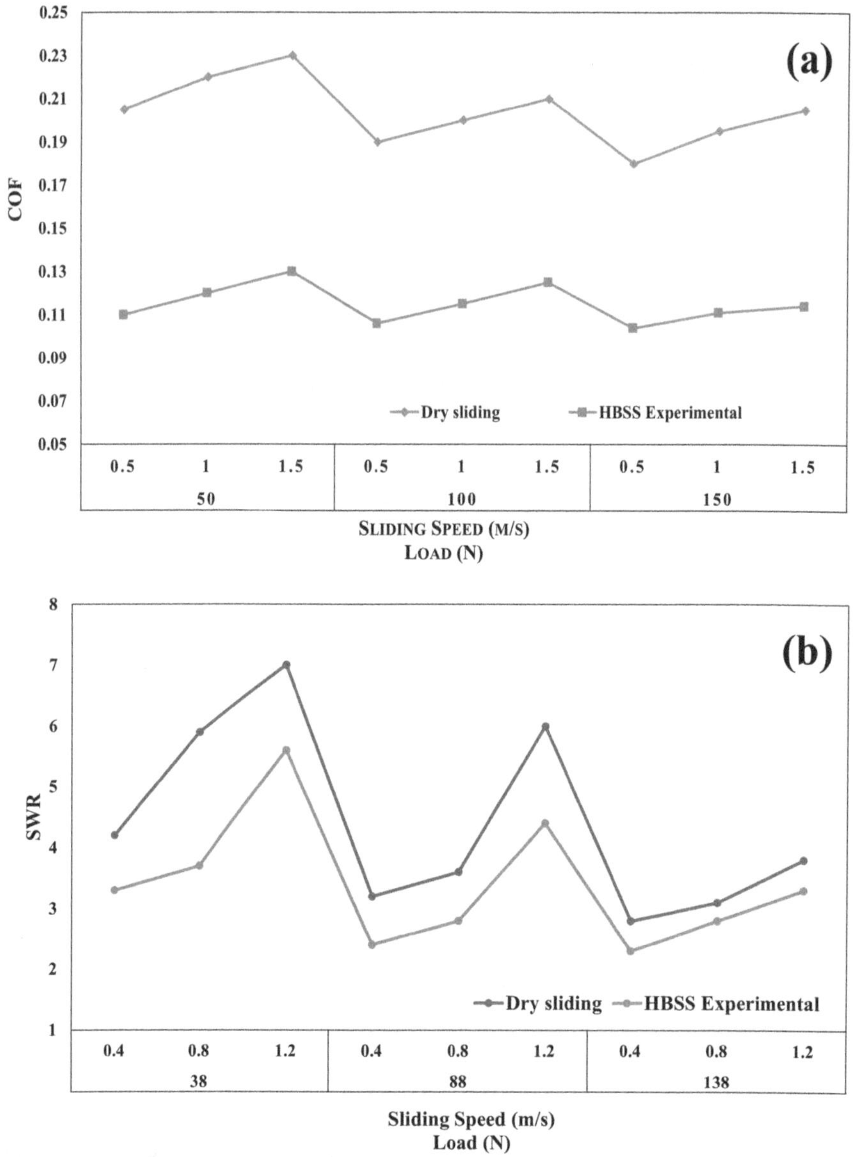

Figure 8.6 UHMW-PE performance under dry sliding and HBSS lubrication (a) COF and (b) SWR.

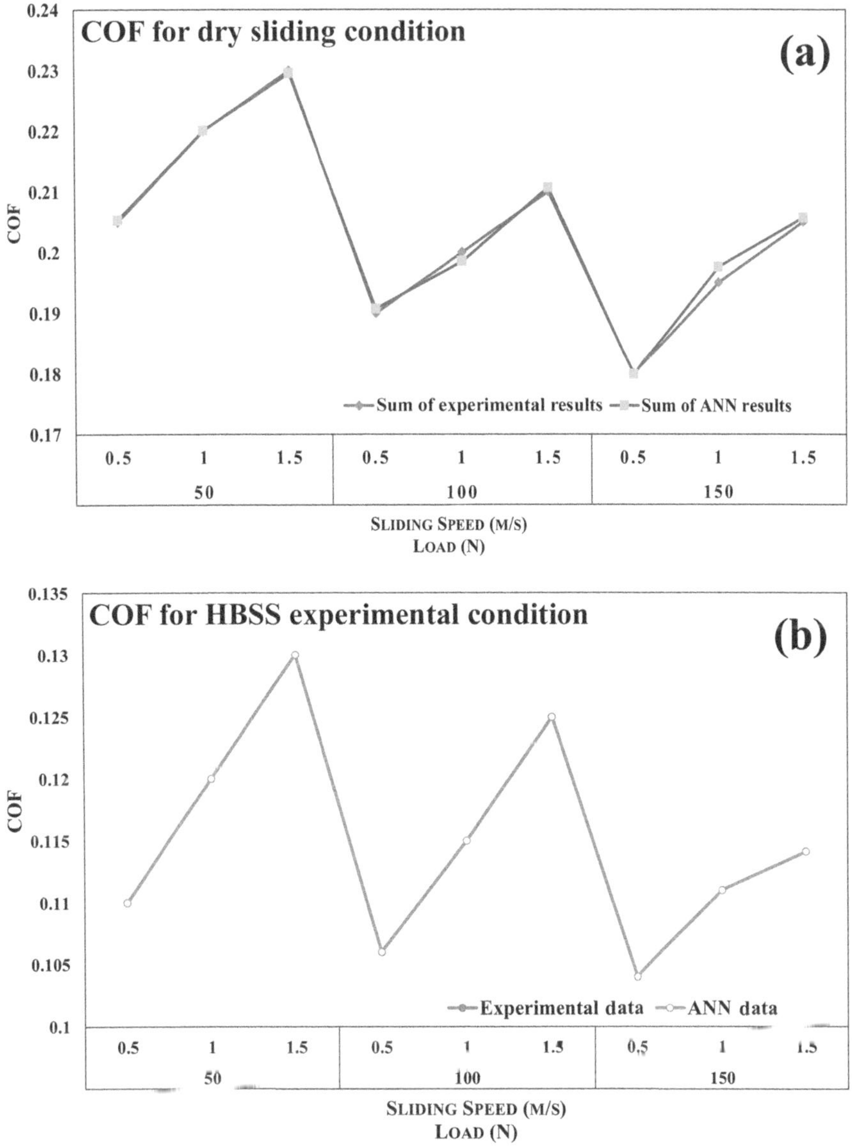

Figure 8.7 UHMW-PE experimental results v/s ANN predictions under (a) COF for dry sliding, (b) COF for HBSS lubrication.

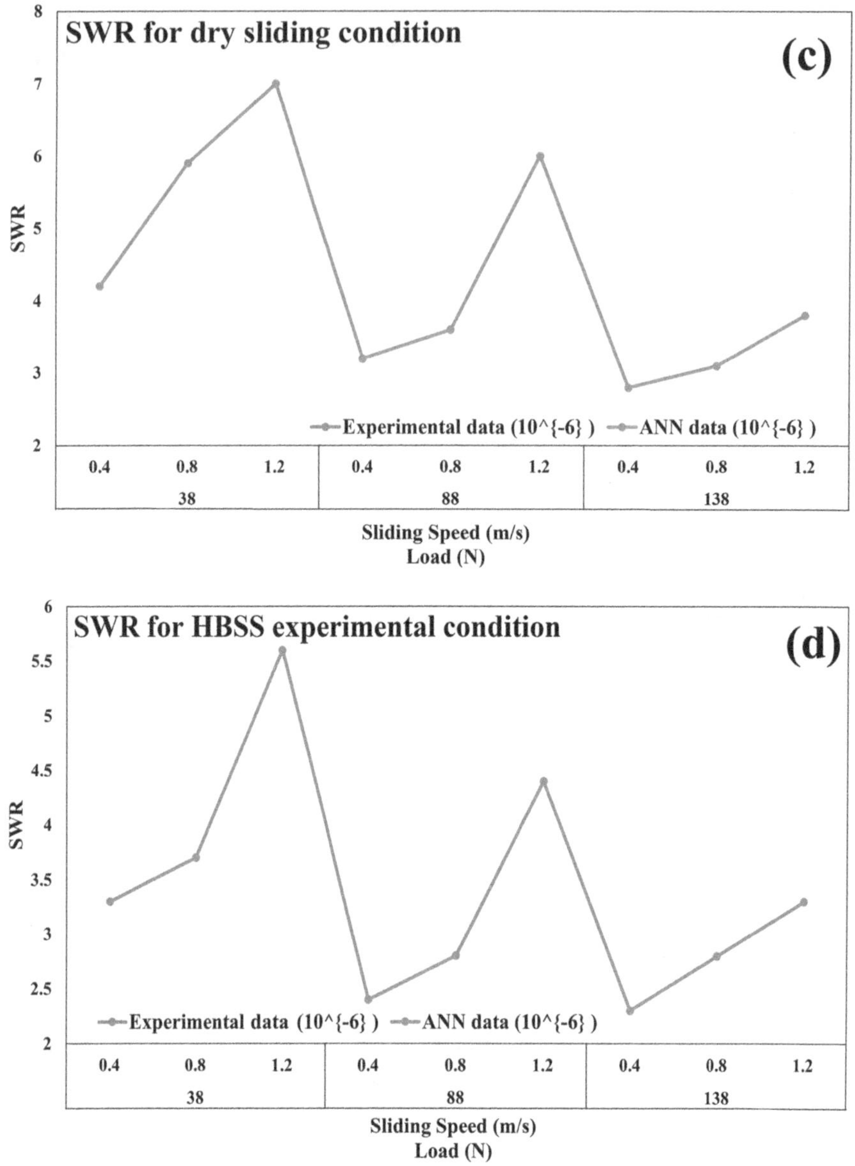

Figure 8.7 (Continued) UHMW-PE experimental results v/s ANN predictions under (c) SWR for dry sliding, and (d) SWR for HBSS lubrication.

Table 8.2 Overview of different ANN models applied in polymer tribology

Material(s)	Selected model(s)	Dataset used for developing model	Model input	Model output	Statistical parameter(s)	Remark(s)	No of hidden layer(s)	Hidden layer neurons	References
Epoxy LY 556 Short glass fibers Blast furnace slag	Back-propagation ANN (4:7:1)	15	Filler content, sliding velocity, sliding distance, normal load	Specific wear rate	Error < 6%	The model is efficient in predicting the dry sliding wear.	1	7	[25]
Epoxy grade (LY556) Castor oil cortex fibers	Regression, layer recurrent ANN, back-propagation feed-forward ANN, backpropagation, cascade forward ANN	36	Sliding distances, fiber lengths, normal loads	Interfacial temperature, wear, COF	Std. Deviation is ±4.5% Error < 1.1%	Purelin transfer function provided the best results for single hidden layer.	3	{9, 13, 9}	[26]
Betelnut fibers Butanox M-60 polyester	ANN (3:2:1)	472	Fiber orientation, applied load, sliding distance	Friction coefficient	SSE < 1%	The combination of Log sigmoid as transfer function and TrainLM as training function produced lowest SSE.	2	$[30]_1[20]_2$	[27]
Chopped strand mat of glass fibers Polyester resin (Revesol P9509)	ANN (4-[40]-1)	7389	Fiber orientation, applied force, rotational speed, sliding duration	Friction coefficient	90% accuracy	The TrainLM function as training function indicted the best result.	1	40	[28]

(Continued)

Table 8.2 (Continued)

Material(s)	*Selected model(s)*	*Dataset used for developing model*	*Model input*	*Model output*	*Statistical parameter(s)*	*Remark(s)*	*No of hidden layer(s)*	*Hidden layer neurons*	*References*
Short carbon fiber TiO_2 Polyphenylene sulfide	Optimal brain surgeon ANN (7-[9]-[3]-1) (wear) & (7-[3]-[1]-1) (COF)	95 (SWR) 90 (COF)	Thermal and mechanical properties of material, Material compositions, Pressure, sliding speed	Coefficient of friction, Specific wear rate	MRE < 0.12 (wear) MRE < 0.78 (COF)	ANN predicted the best composition as PPS +10 vol.% SCF + 3 vol.% TiO_2	2	$[9]_1[3]_2$ (wear) $[3]_1[1]_2$ (COF)	[29]
Short glass fibers, PTFE, graphite fillers, pitch-based carbon fibers, PAN carbon fibers, polyamide 4.6	Backpropagation ANN (9-$[15\text{-}10\text{-}5]_3$-1)	103	Matrix, fiber and PTFE volume, compression modulus, compression strength, hardness, fracture toughness, temperatures, normal force, wear speed	Specific wear rate, coefficient of friction	Error < 1%	The accounting of material composition as input and by increasing dataset helped model for accurate prediction.	3	$[15]_1[10]_2[5]_3$	[30]
Polyamide 4.6, Carbon fibers, PTFE	Backpropagation ANN 9-$[3]_1$-1 (COF) 9-$[12\text{-}6\text{-}3]_3$-1 (wear)	101	Material compositions, temperature, normal force, sliding speed	SWR, COF	MRE < 0.2 (wear) MRE < 0.50 (COF)	One output ANN could provide higher accuracy than multioutput ANN	1(COF) 3 (wear)	3(COF) $[12]_1[6]_2[3]_3$ (wear)	[31]

Graphite, short carbon fiber, PTFE, TiO_2, polyphenylene sulfide	Backpropagation ANNs (7-[3]-[1]-1) (COF), (7-[9]-[3]-1) (wear)	124	Material compositions, sliding speed, applied pressure	SWR, COF	MRE < 0.10 (COF) MRE < 0.55 (wear)	PPS with 3 vol.% TiO_2 and 10 vol.% SCF shows minimum wear	2	$[9]_1[3]_2$ (wear) $[3]_1[1]_2$ (COF)	[32]
Polytetrafluoroethylene, carbon fiber, TiO_2 (titanium dioxide)	Gradient search ANN (7:15:10:5:1)	30–100	Material compositions, sliding speed, load, hardness, compressive strength	Friction coefficient, wear volume loss	Fitting coefficient > 0.9	The prediction accuracy of ANN can be increased by expanding the dataset and with configuration.	3	$[15]_1[10]_2[5]_3$	[33]
PTFE, aramid pulp, PTW, mica, Cu, nano-SiO_2	Monte Carlo-based artificial neural network	18	Material compositions	Specific wear rate, coefficient of friction	RMSE < 0.007 (COF) RMSE < 0.97 (wear)	In comparison with BP-ANN, MC-ANN shows higher accuracy.	1	Not mention	[34]
UHMW-PE, carbon nanotube, graphene oxide, carbon fiber, wollastonite additives	Backpropagation ANN (11:12:1)	125	Type and weight percentages of reinforcements, applied load, Sliding speed	Volume loss	$R^2 > 0.8$	The dataset collected from different research is collectively mentioned.	1	12	[35]
UHMW-PE, MWCNT, graphene	Conjugate gradient backpropagation ANN (7:3:1), (7:5:1)	153	Material compositions, particle size of the nanoparticles	Ultimate tensile strength, Young's modulus	(Young's modulus) $R^2 > 0.93$	The artificial neural network models are used as the objective functions.	1	3, 5	[36]

(Continued)

Table 8.2 (Continued)

Material(s)	*Selected model(s)*	*Dataset used for developing model*	*Model input*	*Model output*	*Statistical parameter(s)*	*Remark(s)*	*No of hidden layer(s)*	*Hidden layer neurons*	*References*
Bisphenol A epoxy resin, GNPs coated carbon fiber, carbon fiber	ANN	45	Material type, temperature, frequency, load	Coefficient of friction	$R^2 > 0.95$	The prediction of the ANN model is in good agreement.	2	$[10]_1[10]_2$	[37]
Carbon fibers, PEEK	BP-ANN (2:7:1), (2:8:1)	30	*pv* factor, contact temperature	Friction coefficient, wear weight loss	Error < 7%	The contact temperature influences the weight loss, and the *pv* factor influences the COF mainly.	1	7, 8	[38]
Polyamide 66	Feed-forward ANN (7-$[20\ 10\ 10]_3$-1)	42	Applied load, load ratio, cyclic loading frequency, number of surface cracks, fluid film thickness	Linear wear rate	Average arithmetic mean error < 3.5 %	The NN configuration and size of input parameters govern the performance of the ANN	3	$[20]_1[10]_2[10]_3$	[39]
Titanium oxide, polyester resin	ANN (4:12:1)	9	Sliding velocity, normal load, filler content, sliding distance	Sp. wear rate	Error < 8%	Sliding wear resistance of polyester matrix composite improves with TiO_2 as filler	1	12	[40]

8.4 CASE STUDY

8.4.1 Dataset description

In the dataset employed for this study, a total of 22,631 samples were analyzed to explore the application of ANNs in predicting wear and COF in polymer tribology. The dataset features a variety of parameters significant for the tribological assessment. These parameters include the 'Composition' of the polymers, presented both qualitatively and as percentage ('Composition (%)'), which provides insights into the material properties influencing tribological behavior. Mechanical stress conditions are represented by 'Load (N)' and 'Speed (rpm)', which describe the operational environment of the polymer materials under study. Additionally, 'SD (mm)' denotes the sliding distance, indicating the extent of contact and movement against counterfaces, and 'Time (Seconds)' reflects the duration of each test, correlating with the wear process duration.

The output variables of interest, 'Wear (μm)' and 'COF', are critical in evaluating the performance and durability of polymer materials under specified conditions. 'Wear (μm)' quantifies the material loss due to frictional forces, and 'COF' measures the resistance encountered when one surface slides over another. The inclusion of these outputs allows for a comprehensive analysis of tribological behavior through ANN modeling, facilitating a deeper understanding of the relationships between material composition, operational conditions, and tribological outcomes. This dataset is meticulously detailed in Table 8.3 of the study, providing a robust foundation for applying ANN techniques to predict and analyze wear and COF in polymer-based materials. The frequency distribution histogram plot of the dataset variables is presented in Figure 8.8. Furthermore, the Pearson correlation matrix is depicted in Figure 8.9 showing variables' linear correlations.

8.5 RESULTS AND DISCUSSIONS

The ANN model demonstrated varying degrees of accuracy in predicting the two key outputs of interest in polymer tribology, as shown in Table 8.3: wear and the COF. The model's performance in predicting wear was notably strong, as indicated by the R^2 value of 0.837. This suggests that a high level of variance in wear measurements was captured by the ANN, highlighting its effectiveness in modeling this aspect of tribological behavior. In contrast, the model's ability to predict COF was considerably less successful, as reflected by a significantly negative R^2 value of −17.722, indicating that the model predictions deviated extensively from the actual COF values and might even be worse than a simple average.

The MSE and RMSE for wear were 370.600 and 19.251, respectively, suggesting that while the model captures the general trends in wear,

Table 8.3 Dataset descriptive analysis

Name	*Mean*	*Standard deviation*	*Minimum*	*Median*	*Maximum*	*Variance*	*Skewness*	*Kurtosis*	*1st quartile (Q1)*	*3rd quartile (Q3)*	*Interquartile range (Q3–Q1)*
Composition	21.31943	13.58896	0	30	35	184.6598	−0.6608	−1.13838	0	35	35
Load	29.21586	7.92833	19.62	29.43	39.24	62.85845	0.0397	−1.46797	19.62	39.24	19.62
Speed	982.767	405.6211	500	1000	1500	164528.5	0.06298	−1.47789	500	1500	1000
SD	1501.259	414.5878	1000	1500	2000	171883.1	−0.00469	−1.54552	1000	2000	1000
Time	299.3088	172.2659	0.906	299.204	634.375	29675.52	0.00467	−1.19248	150.078	448.343	298.265
Wear	63.13144	47.51591	0	51.44	223.38	2257.761	1.0824	0.83531	29.29	88.55	59.26
COF	0.49001	0.21884	−0.11162	0.44343	1.20846	0.04789	0.8928	0.68978	0.34217	0.61417	0.272

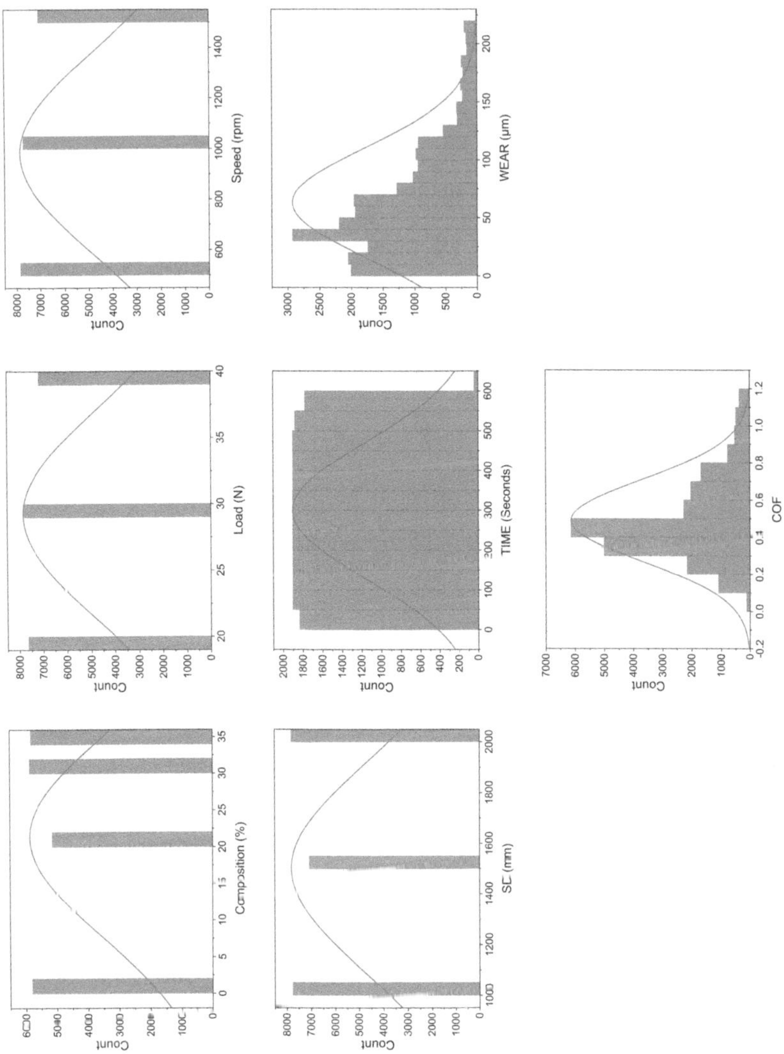

Figure 8.8 Frequency distribution histogram plot.

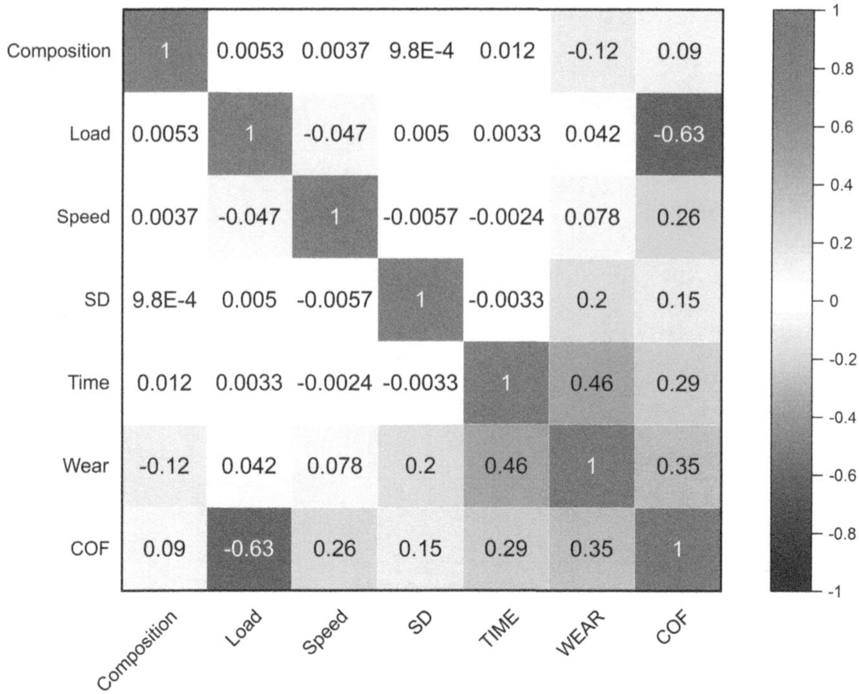

Figure 8.9 Pearson correlation matrix of the variables.

individual predictions can deviate significantly from the true values. The MSE and RMSE for COF were alarmingly high at 0.878 and 0.937, respectively, further affirming the challenges faced by the ANN in accurately predicting frictional behavior under varying conditions.

The MAE, which provides a clearer measure of average prediction error magnitude, stood at 13.968 for wear and 0.884 for COF as presented in Table 8.4. These values reinforce the disparity in prediction accuracy between wear and COF, with the model showing a reasonable predictive capability for wear but struggling significantly with COF.

The ANN model's robust performance in predicting wear suggests that the input features related to composition, load, speed, sliding distance, and operational time are well-correlated with wear outcomes and are effectively utilized by the ANN. This is critical for tribological studies, where predicting material wear under diverse conditions can significantly impact material selection and design.

However, the model's poor performance in predicting COF calls for a critical evaluation. The negative R^2 value is a stark indicator that the model predictions deviate from the observed data to a great extent, which could be

Table 8.4 ANN model performance to predict outputs

Model	*R^2 for output 1*	*R^2 for output 2*	*MSE y1*	*MSE y2*	*RMSE y1*	*RMSE y2*	*MAE y1*	*MAE y2*
ANN	0.837	−17.722	370.600	0.878	19.251	0.937	13.968	0.884

due to several factors. These might include insufficient or noisy data concerning factors that significantly influence COF, or perhaps the ANN architecture is not suitably configured to capture the complex dynamics affecting COF. This aspect of friction is influenced by a multitude of factors, including surface interactions, temperature variations, and material microstructure changes under stress, which may not be adequately captured by the available dataset.

The results presented in Figures 8.10 and 8.11, which illustrate actual versus predicted plots for both outputs, visually confirm these interpretations. The plots likely show a good alignment of predicted versus actual values for wear, with deviations becoming starkly apparent in the COF predictions.

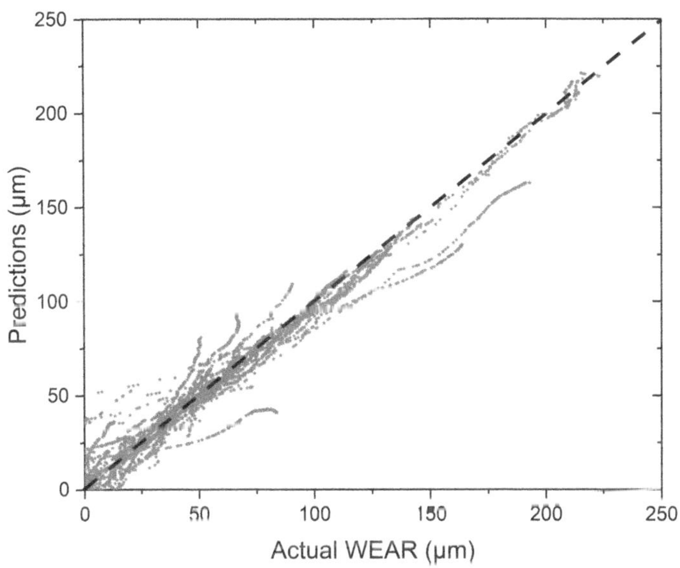

Figure 8.10 Actual versus prediction plot of WEAR.

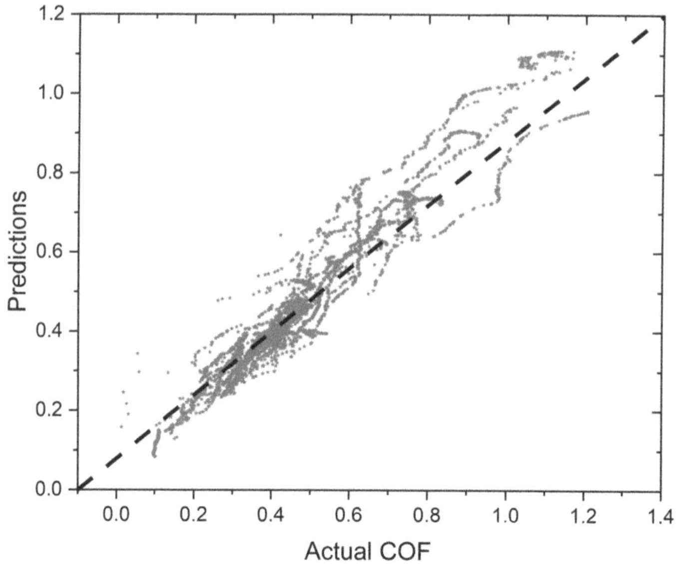

Figure 8.11 Actual versus prediction plot of COF.

8.6 FEATURE IMPORTANCE ANALYSIS

The analysis of feature importance for predicting wear and the COF in polymer tribology provides insightful details into the relative significance of various operational and material properties processed by ANN, as shown in Figures 8.12 and 8.13.

8.6.1 Wear prediction

For wear prediction, the most influential feature is 'TIME (Seconds)', with an importance score of 0.2425. This indicates that the duration of the tribological test plays a crucial role in determining wear, possibly because longer test durations allow for more extensive material interactions and degradation processes. Following closely is 'Speed (rpm)' with an importance score of 0.2184, suggesting that the relative velocity between interacting surfaces significantly impacts the wear rate, likely due to increased thermal and mechanical stresses at higher speeds.

'Load (N)' also shows substantial importance with a score of 0.2149, underscoring the impact of mechanical pressure on wear. Higher loads typically result in increased contact pressure, enhancing wear through more aggressive mechanical interactions. 'Composition (%)', which has an importance score of 0.1900 reflects the role of material composition in wear characteristics. Different compositions can drastically alter material properties such as hardness and resilience, which in turn affect wear outcomes.

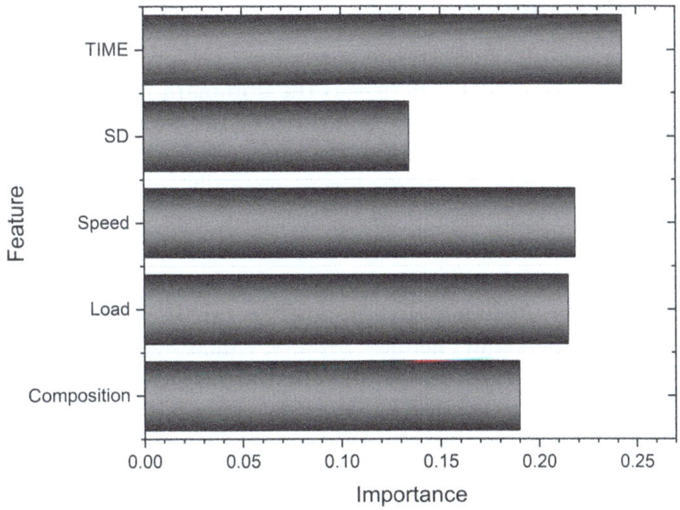

Figure 8.12 Feature importance for predicting WEAR.

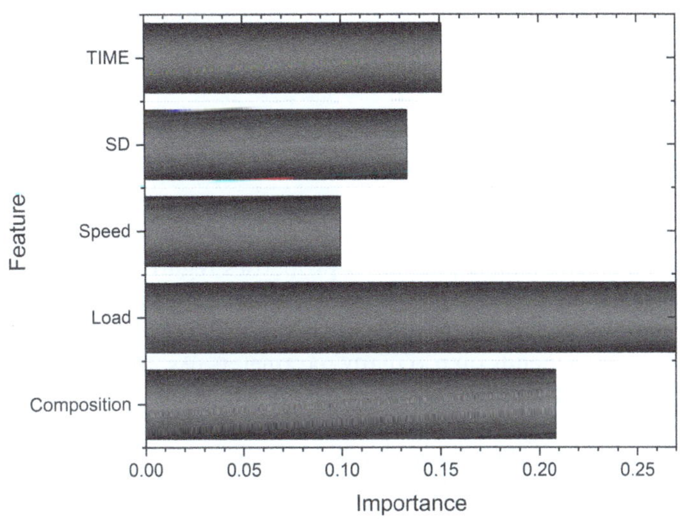

Figure 8.13 Feature importance for predicting COF.

Lastly, 'SD (mm)' or sliding distance has an importance score of 0.1342. Being significant, it has the least impact among the analyzed features. This suggests that while the extent of the sliding interaction affects wear, other dynamic factors such as load and speed are more critical under the conditions tested.

8.6.2 COF prediction

In contrast, the feature importance scores for COF prediction highlight different priorities. 'Load (N)' emerges as the dominant factor with an importance score of 0.4076, indicating that the pressure exerted on the materials is paramount in influencing frictional behavior. This aligns with tribological principles where increased normal forces generally enhance frictional resistance between surfaces.

'Composition (%)' follows with an importance of 0.2085, reinforcing the significance of material makeup in determining how surfaces interact under frictional conditions. Changes in composition can alter surface characteristics such as roughness and chemical affinity, which significantly affect COF.

'TIME (Seconds)' has a moderate importance of 0.1509, suggesting a temporal component to friction behavior, possibly related to wear-in effects or changes in surface conditions over time. 'SD (mm)' and 'Speed (rpm)' have lower importance scores of 0.1334 and 0.0996, respectively. The relatively lower importance of speed for COF, compared to its role in wear prediction, might indicate that while speed alters wear rates through thermal and mechanical stress, its direct influence on COF is less pronounced under the tested conditions.

This feature importance analysis not only aids in understanding the dynamics of wear and COF in polymer tribology but also assists in refining the ANN models for better predictive accuracy by focusing on the most impactful parameters.

8.7 TRENDS AND CONCLUDING REMARKS

ANN proves an efficient statistical tool for predicting the performance of the material by the wear and friction properties regarding the input parameters. The architecture of ANN is quite complex and is able to predict the highly complex problems in nature, such as the tribological properties of material, which depends on too many parameters. As presented in Table 8.2, the trends of implementing the varieties of ML tools in combination with data pre-processing and analyzing by increasing the training and testing datasets and by including more numbers of parameters as input give highly accurate predictions for SWR and COF. The understanding becomes clearer for the behavior of the material in complex environments where too many parameters affect the material properties with the help of ANN-like neural networks. The use of neural networks in association with ML models and other statistical tools opens the door to attempting more challenging problems, which were not possible earlier through conventional statistical tools. The process timing in solving the complex problems and human brain limitations to understand the higher-dimensional problems are possible reasons why neural networks have become much popular nowadays. As new attempts create new challenges in research communities, efforts are made to make ANN more efficient and powerful with less computational time. We encourage research communities to develop new robust and efficient methods for the tribology domain, which helps us to move toward an energy-efficient and greener era by optimizing the tribological properties of new as well as conventional materials for different applications.

REFERENCES

1. J. O. Kephart and D. M. Chess, "The vision of autonomic computing," *Computer (Long Beach Calif)*, vol. 36, no. 1, pp. 41–50, 2003, doi: 10.1109/MC.2003.1160055
2. A. Antoniadis, S. Lambert-Lacroix, and F. Leblanc, "Effective dimension reduction methods for tumor classification using gene expression data," *Bioinformatics*, vol. 19, no. 5, pp. 563–570, March. 2003, doi: 10.1093/BIOINFORMATICS/BTG062
3. R. W. Carpick and M. Salmeron, "Scratching the surface: Fundamental investigations of tribology with atomic force microscopy," *Chem Rev*, vol. 97, no. 4, pp. 1163–1194, 1997, doi: 10.1021/CR960068Q/ASSET/CR960068Q.FP.PNG_V03.
4. A. Kordijazi, T. Zhao, J. Zhang, K. Alrfou, and P. Rohatgi, "A review of application of machine learning in design, synthesis, and characterization of metal matrix composites: Current status and emerging applications," *JOM*, vol. 73, no. 7, pp. 2060–2074, July 2021, doi: 10.1007/S11837-021-04701-2
5. H. Chan et al., "Machine learning classical interatomic potentials for molecular dynamics from first-principles training data," *J Phys Chem C*, vol. 123, no. 12, pp. 6941–6957, March 2019, doi: 10.1021/ACS.JPCC.8B09917
6. B. Eren, M. A. Guvenc, and S. Mistikoglu, "Artificial intelligence applications for friction stir welding: A review," *Met Mater Int*, vol. 27, no. 2, pp. 193–219, February 2021, doi: 10.1007/S12540-020-00854-Y
7. Z. Wang et al., "The role of machine learning to boost the bioenergy and biofuels conversion," *Bioresour Technol*, vol. 343, p. 126099, January 2022, doi: 10.1016/J.BIORTECH.2021.126099
8. A. Géron, *Hands-on machine learning with Scikit-Learn, Keras, and TensorFlow*. O'Reilly Media, Inc., 2022.
9. I. A. Basheer and M. Hajmeer, "Artificial neural networks: Fundamentals, computing, design, and application," *J Microbiol Meth*, vol. 43, no. 1, pp. 3–31, December 2000, doi: 10.1016/S0167-7012(00)00201-3
10. O. Mohamed, M. Kewalramani, M. Ati, and W. Al Hawat, "Application of ANN for prediction of chloride penetration resistance and concrete compressive strength," *Materialia (Oxf)*, vol. 17, p. 101123, 2021, doi: 10.1016/j.mtla.2021.101123
11. L. Huang, J. Chen, and X. Tan, "BP-ANN based bond strength prediction for FRP reinforced concrete at high temperature," *Eng Struct*, vol. 257, p. 114026, April 2022, doi: 10.1016/J.ENGSTRUCT.2022.114026
12. J. H. Haido, "Prediction of the shear strength of RC beam-column joints using new ANN formulations," *Structures*, vol. 38, no. February, pp. 1191–1209, 2022, doi: 10.1016/j.istruc.2022.02.046

13. M. W. Ahmad, M. Mourshed, and Y. Rezgui, "Trees vs neurons: Comparison between random forest and ANN for high-resolution prediction of building energy consumption," *Energy Build*, vol. 147, pp. 77–89, July 2017, doi: 10.1016/J.ENBUILD.2017.04.038
14. E. M. Golafshani, A. Behnood, and M. Arashpour, "Predicting the compressive strength of normal and High-Performance Concretes using ANN and ANFIS hybridized with Grey Wolf Optimizer," *Constr Build Mater*, vol. 232, p. 117266, January 2020, doi: 10.1016/J.CONBUILDMAT.2019.117266
15. R. K. Tipu, V. R. Panchal, and K. S. Pandya, "Prediction of concrete properties using machine learning algorithm," *J Phys Conf Ser*, vol. 2273, no. 1, p. 012016, May 2022, doi: 10.1088/1742-6596/2273/1/012016
16. R. Kumar Tipu, V. R. Panchal, and K. S. Pandya, "An ensemble approach to improve BPNN model precision for predicting compressive strength of high-performance concrete," *Structures*, vol. 45, pp. 500–508, November 2022, doi: 10.1016/J.ISTRUC.2022.09.046
17. J. V. Tu, "Advantages and disadvantages of using artificial neural networks versus logistic regression for predicting medical outcomes," *J Clin Epidemiol*, vol. 49, no. 11, pp. 1225–1231, November 1996, doi: 10.1016/S0895-4356(96)00002-9
18. S. Jabin, "Stock market prediction using feed-forward artificial neural network," *Int J Comput Appl*, vol. 99, no. 9, pp. 4–8, 2014, doi: 10.5120/17399-7959
19. A. Iqbal and Shabib A, "A feed-forward and pattern recognition ANN model for network intrusion detection," *Int J Comput Netw Inf Secur*, vol. 11, no. 4, 2019, doi: 10.5815/ijcnis.2019.04.03
20. D. A. Cirovic, "Feed-forward artificial neural networks: Applications to spectroscopy," *TrAC Trends Anal Chem*, vol. 16, no. 3, pp. 148–155, March 1997, doi: 10.1016/S0165-9936(97)00007-1
21. F. Anctil, C. Perrin, and V. Andréassian, "Impact of the length of observed records on the performance of ANN and of conceptual parsimonious rainfall-runoff forecasting models," *Environ Modell Softw*, vol. 19, no. 4, pp. 357–368, April 2004, doi: 10.1016/S1364-8152(03)00135-X
22. A. P. Piotrowski and J. J. Napiorkowski, "A comparison of methods to avoid overfitting in neural networks training in the case of catchment runoff modelling," *J Hydrol (Amst)*, vol. 476, pp. 97–111, January 2013, doi: 10.1016/J.JHYDROL.2012.10.019
23. N. Sharma, S. Kumar, and K. K. Singh, "Taguchi's DOE and artificial neural network analysis for the prediction of tribological performance of graphene nano-platelets filled glass fiber reinforced epoxy composites under the dry sliding condition," *Tribol Int*, vol. 172, August 2022, doi: 10.1016/j.triboint.2022.107580
24. K. Ermiş and H. Ünal, "Tribological behavior of ultra-high molecular weight polyethylene polymer with artificial neural network modeling," *Kocaeli J Sci Eng*, September 2021, doi: 10.34088/kojose.961118
25. P. K. Padhi and A. Satapathy, "Analysis of sliding wear characteristics of BFS filled composites using an experimental design approach integrated with ANN," *Tribol Trans*, vol. 56, no. 5, pp. 789–796, September 2013, doi: 10.1080/10402004.2013.798448
26. R. Egala, G. V. Jagadeesh, and S. G. Setti, "Experimental investigation and prediction of tribological behavior of unidirectional short castor oil fiber reinforced epoxy composites," *Friction*, vol. 9, no. 2, pp. 250–272, April 2021, doi: 10.1007/s40544-019-0332-0

27. U. Nirmal, "Prediction of friction coefficient of treated betelnut fibre reinforced polyester (T-BFRP) composite using artificial neural networks," *Tribol Int*, vol. 43, no. 8, pp. 1417–1429, August 2010, doi: 10.1016/j.triboint.2010.01.013
28. T. Nasir, B. F. Yousif, S. McWilliam, N. D. Salih, and L. T. Hui, "An artificial neural network for prediction of the friction coefficient of multi-layer polymeric composites in three different orientations," *Proc Inst Mech Eng, Part C: J Mechan Eng Sci*, February 2010, vol. 224, no. 2, pp. 419–429. doi: 10.1243/09544062JMES1677
29. L. A. Gyurova, P. Miniño-Justel, and A. K. Schlarb, "Modeling the sliding wear and friction properties of polyphenylene sulfide composites using artificial neural networks," *Wear*, vol. 268, no. 5–6, pp. 708–714, February 2010, doi: 10.1016/j.wear.2009.11.008
30. Z. Zhang, K. Friedrich, and K. Velten, "Prediction on tribological properties of short fibre composites using artificial neural networks," *Wear*, vol. 252, no. 7–8, pp. 668–675, 2002, doi: 10.1016/S0043-1648(02)00023-6
31. Z. Jiang, L. Gyurova, Z. Zhang, K. Friedrich, and A. K. Schlarb, "Neural network based prediction on mechanical and wear properties of short fibers reinforced polyamide composites," *Mater Des*, vol. 29, no. 3, pp. 628–637, 2008, doi: 10.1016/j.matdes.2007.02.008
32. L. A. Gyurova and K. Friedrich, "Artificial neural networks for predicting sliding friction and wear properties of polyphenylene sulfide composites," *Tribol Int*, vol. 44, no. 5, pp. 603–609, May 2011, doi: 10.1016/j.triboint.2010.12.011
33. J. Zhu, Y. Shi, X. Feng, H. Wang, and X. Lu, "Prediction on tribological properties of carbon fiber and TiO_2 synergistic reinforced polytetrafluoroethylene composites with artificial neural networks," *Mater Des*, vol. 30, no. 4, pp. 1042–1049, April 2009, doi: 10.1016/j.matdes.2008.06.045
34. S. Li et al., "Tribological behavior prediction of friction materials for ultrasonic motors using Monte Carlo-based artificial neural network," *J Appl Polym Sci*, vol. 136, no. 10, p. 47157, March 2019, doi: 10.1002/APP.47157
35. H. I. Kurt and M. Oduncuoglu, "Application of a neural network model for prediction of wear properties of ultrahigh molecular weight polyethylene composites," *Int J Polym Sci*, vol. 2015, 2015, doi: 10.1155/2015/315710
36. A. Vinoth and S. Datta, "Design of the ultrahigh molecular weight polyethylene composites with multiple nanoparticles: An artificial intelligence approach," *J Compos Mater*, vol. 54, no. 2, pp. 179–192, January 2020, doi: 10.1177/0021998319859924
37. M. Singh, S. Dodla, R. K. Gautam, and V. K. Srivastava, "Effect of load, sliding frequency, and temperature on tribological properties of graphene nanoplatelets coated carbon fiber reinforced polymer composites," *J Compos Mater*, 2022, doi: 10.1177/00219983221140205
38. X. LiuJie, J. P. Davim, and R. Cardoso, "Prediction on tribological behaviour of composite PEEK-CF30 using artificial neural networks," *J Mater Process Technol*, vol. 189, no. 1–3, pp. 374–378, July 2007, doi: 10.1016/j.jmatprotec.2007.02.019
39. A. Abdelbary, M. N. Abouelwafa, I. M. el Fahham, and A. H. Hamdy, "Modeling the wear of Polyamide 66 using artificial neural network," *Mater Des*, vol. 41, pp. 460–469, October 2012, doi: 10.1016/j.matdes.2012.05.013
40. A. Satapathy, D. P. Tarkes, and N. B. Nayak, "Wear response prediction of TiO_2-polyester composites using neural networks," *International Journal of Plastics Technology*, vol. 14, no. 1 SUPPL, 2010, doi: 10.1007/s12588-010-0004-4

Chapter 9

Potential of natural bast fibers as reinforcement in polymer matrix composites

Gokulkumar Sivanantham, Aravindh Murugan, Gurupranes Sivaraj Vijaya, and Karthi Nadarajan

9.1 INTRODUCTION

Bast fibers are an excellent, environmentally friendly product, as their raw materials are renewable and the production processes consume less energy. It can also be disposed of through composting, which further reduces its impact on the environment. It also provides considerable economic returns and, when used as reinforcement in PMCs, yields low-cost materials. These PMCs are lightweight, with mechanical properties that are adequate [1–3].

While so many positive factors are associated with the fibers, research in bast fiber-based PMCs is still in its development stage, and more efforts are required for understanding the full capacity of the bast fibers. Optimization of the processing conditions to achieve finer dispersion together with the development of new processing techniques to enhance the mechanical properties is a major area of research activity. The shift to sustainable materials has led to the formulation of bast-fiber-based PMCs, helpful in achieving environmental and economic benefits [4–7]. These PMCs are expected to replace conventional materials utilized in PMCs, thereby enhancing sustainability across various industries.

The manufacturing technique primarily depends on the final product and its application. A few common manufacturing techniques include a hand layup method, fiber spinning, compression molding, and injection molding, among others. According to [8–11], the hand layup technique is one of the simplest and low-cost methods of fabricating PMCs parts. This method is the laying of fibers manually in a mold, where the resin is applied by hand. The technique generally results in large parts or parts with simple geometry. Such fibers produced in this manner can be allowed to reinforce in PMCs similar to other natural fibers. Compression molding is another approach common in the manufacture of PMCs parts: placing resin-intermixed fibers inside a mold, followed by pressure and heat to consolidate the material. This process generally produces high-volume parts of consistent quality [12]. Injection molding is a technique of manufacturing PMC parts through the injection of resin into a mold. This technique is very common in the manufacture of parts that have complicated shapes and is useful for making

 DOI: 10.1201/9781003527275-9

high-volume parts of consistent quality [13]. Various techniques are used in making natural fiber-reinforced polymer composites each having its own merits and demerits. The choice of the technique always depends on the final product and intended application. The best method is determined by careful consideration of the requirements of an application.

Mechanical properties are very vital when material selection is done for specific applications. Reasons contributing to this factor include the type of bast fiber, fiber content, fiber orientation, and resin type. The chemical treatment of the fibers can also cause a change in the mechanical properties of the polymer matrix composite. It generally performs better than glass fiber-reinforced polymer composites but is inferior to carbon fiber-reinforced composites [14–20].

Apart from the mechanical properties, the thermal properties are very significant, depending on the type of fiber and resin used, including fiber orientation and content. They were found to have higher thermal conductivity and the coefficient of thermal expansion than glass FRC and lower than carbon FRCs. Therefore, considering the selection of materials for a particular application, both mechanical properties and thermal properties will be of utmost importance. In this context, bast-fiber-reinforced PMCs show a proper balance between mechanical and thermal properties that is suitable for many applications.

Thus, the motivation for this review lies in exploring how bast-fiber reinforced polymer composites can serve as a class of sustainable alternatives with promising mechanical and thermal properties. This idea is driven by the dire need to address environmental concerns coupled with the economic benefits derivable from such materials across many industries. The contribution of this study in this respect would be an attempt at bridging the existing knowledge gap through the exploration and optimization of bast fibers in PMC applications for the greater good of promoting sustainability.

9.1.1 Bast fiber as reinforcing material

Fibers originating from the phloem or inner bark of plants such as flax, hemp, jute, ramie, and banana are, in general, termed bast fibers. Bast fibers are an exciting new family of natural polymers that have recently been increasingly explored as reinforcements in polymeric materials. The bast fibers have a number of advantages over conventional reinforcement materials such as glass fibers and carbon fibers [21–24]. In addition, low-density bast fibers are low-cost materials with satisfactory mechanical properties involving tensile strength and modulus of elasticity. Therefore, the mechanical properties of these NFRCs based on bast fibers can be improved by chemical treatment and optimization of the fiber content and orientation. Overall, bast fibers are a more attractive alternative to traditional reinforcement materials. Nevertheless, since the inner bark of several plants exhibits a hydrophilic characteristic, there exists inadequate compatibility between the fiber and matrix. This results in a disparity in load distribution between the fiber and the matrix. Consequently, chemical treatment is essential to enhance the hydrophobicity of the fibers and augment compatibility with the matrix.

9.1.2 Chemical analysis of bast fiber

Mechanical and thermal properties, as well as stability regarding environmental factors such as moisture, heat, and UV radiation, depend on the chemical composition of bast fibers. For instance, the presence of lignin in the fibers contributes to their resistance to moisture, while hemicellulose increases their flexibility. Bast fibers usually contain a high content of cellulose in the range from 60% to 75% [25]. Cellulose is the main structural component of plant cell walls and provides fibers with their strength and stiffness. Other components, in addition to cellulose, that are generally considered to make up bast fibers are hemicellulose, lignin, pectin, and ash. Hemicellulose, a polysaccharide combined with cellulose, forms a three-dimensional network structure in the plant cell wall and confers flexibility to the fiber. While lignin is a complex polymer giving rigidity and strength to the plant cell walls, pectin is a polysaccharide responsible for maintaining the adhesion among plant cells. The ash content of bast fibers is comparatively less than 2%. The microfibril angle is also another important entity in regulating and controlling the strength and stiffness of the plant fibers, including tensile strength. Typically, high crystallinity index and very small-sized MFs lead to excellent tensile strength in these fibers due to parallel alignment of MFs along the fiber axis that assists in the best reinforcement and thus results in an enhancement of the overall strength of the fibers.

9.2 BAST-FIBER-BASED NATURAL-REINFORCED COMPOSITE

Over the past ten years, the properties of bast-fiber-based NFRC have been improved by diverse methodologies ranging from chemical treatment to the incorporation of filler materials. New processing techniques are being developed that can produce high-performance polymer matrix composites with enhanced mechanical and thermal properties. Among the bast fibers, five different species have been considered more particularly in reinforcement: banana, flax, hemp, jute, and ramie. All these are illustrated in Figure 9.1. All of them present peculiar characteristics, which render them more or less suitable for specific applications; a short overview of their properties is given in the next section. As a general feature, bast fibers present high cellulose content, but some varieties, such as ramie and hemp, attain particularly high values.

Figure 9.1 Bast fibers.

9.2.1 Flax-fiber-based NFRCs

Flax is a natural fiber that is taken out of the stem of the flax plant. It is an extremely valuable crop, very friendly to the environment, and has been used since many centuries due to its fibers and even nowadays because of its wide variety of properties and benefits [26]. It has several important advantages, such as high specific strength, which makes it feasible for application in lightweight structures and applications where weight reduction is crucial. In addition, they show good thermal and acoustic properties while being biodegradable, hence establishing their environment-friendly nature [27–30].

In fact, flax-reinforced polymer matrix composites show very good dimensional stability, resistance against moisture and UV light, and thus are suitable for outdoor application. This is also coupled with a low thermal expansion coefficient and hence suitable for high-temperature applications. Processing of flax-fiber-based PMCs is comparatively easy and can be done using conventional techniques, that is, either by injection molding or compression molding. However, in the treatment process for flax-fiber-based PMCs, problems arise because of the brittle nature of flax fibers, which makes homogenous distribution in a matrix difficult.

Moreover, increased stack ups of flax fibers showed improvements in both tensile and flexural strengths [31]. Flax fibers have also been studied to investigate the significant influence of surface modifications. The results presented good insight into and showed the capability of such materials in various applications. Similarly, El-Sakhawy and Hassan [19] reported that the interfacial shear strength of these PMCs considerably reduced at 70% and up to 90% relative humidity. That means control of the level of relative humidity for the fabrication of NFRCs must be observed and thus requires air-conditioning and dehumidification facilities to maintain consistency in reliable performance.

Giuliani et al. [32] studied the long-term loading behavior of flax-fiber-reinforced PMCs. Among the important mechanical properties, describing the deformation of a material under a load as time progresses, are creep and stress relaxation. The present study aims to provide insight into the behavior of creep and stress relaxation of flax-fiber-based NFRCs and assesses the feasibility of the material in various applications. The experimental results showed that flax-fiber-based NFRCs possess improved properties, which will enable the designer to fabricate these kinds of PMCs.

Rajesh et al. [33] studied the thermal behavior of flax-fiber/ramie-fiber-based hybrid NFRCs by performing TG analysis to investigate the weight losses at different temperatures. More significantly, the sample showed less weight loss in the first stage, while the weight losses were more in the second and third stages. The maximum value obtained was 630°C, showing that sample H had a higher heat deflection temperature due to its optimized composition, while sample C had the lowest values.

Based on similar concepts, Paulo et al. [34] demonstrated that the infill density of the flax-reinforced polymer composite significantly influenced strength and stiffness, whereas the surface chemical treatment had no effect. They further showed that the in-plane strength and bending strength of the flax fibers are higher. The maximum values of tensile and flexural stresses were 50 and 73 MPa, respectively, which means the flax reinforcement can enhance the mechanical strength of the PMCs.

Zhu et al. [35] showed the enhancement of the mechanical properties of flax-fiber/polystyrene-based NFRCs and highlighted the promising potential of using such PMCs in various applications. The use of waste material, in this case, expanded polystyrene, will also help reduce waste and contribute to sustainability; hence, this is a promising research area. The results of the present work, such as maximum tensile, flexural, and impact strengths at 30% fiber weight fraction, might be useful in further development and optimization of such PMCs.

Therefore, Samant et al. [36] drew the conclusion that surface treatments to flax fiber improve the mechanical strength of natural rubber-based NFRCs. It was observed that the optimum condition for fiber length and loading was 1.25 cm and 40%, respectively, through tensile analysis. Treatment of the fiber surface with NaOH and laccase significantly improved the properties, such as tensile and tear strength, and also the interfacial contact between the fiber and natural rubber matrix. These results indicate that the reinforcement surface treatment of natural fibers can significantly improve the properties of natural rubber PMCs, hence making them suitable for various applications.

9.2.2 Jute-fiber-based NFRCs

Jute is a member of the family *Tiliaceae*. The botanical name is *Corchorus capsularis*. Jute fibers are obtained by chopping off the plant, followed by submerging in water, and then eventually separating the inner stem from the outside fiber, through a process called retting. Siddiquee and Helali concluded that the jute polymer PMCs had higher rates of degradation under natural weather conditions. During biodegradation, the fiber length and fiber ratio at 1%, 3%, 5%, 10%, and 15% had an effect on the rate of degradation [37–39].

In this context, Rezghi Maleki et al. [40] have identified key parameters for selecting appropriate cutting conditions and drill bit types in the drilling process of jute fiber-based natural fiber-reinforced hybrid composites (NFRHCs) and have also highlighted the possibility of research extension. The obtained results provide important information to industry professionals and researchers concerned with the development and production of PMCs. It provided greater effectiveness and quality in the drilling process of PMCs developed through pointing out the optimum type of drill bit and cutting parameters.

A recent work by Sajin et al. [41] has presented the effect of fiber length on the mechanical strengths of jute-fiber-based NFRHCs. In this work, PMCs were molded for various fiber lengths such as 5, 10, 15, 20 and 25 mm and then characterized for tensile, bending, and energy absorption according to ASTM. The optimum mechanical properties for the developed PMCs were obtained when the fiber length was 5 mm.

Flammability resistance in the case of woven jute-fiber-based NFRHCs was studied by Balan et al. [42], incorporating fly ash in varying proportions. The addition of fly ash was found useful for minimizing flammability. The best result obtained was with the incorporation of 5% by weight of jute fiber and 15% by weight of fly ash treated with sodium hydroxide. The study then compares the results by applying different machine learning regression models, among which multilayer perception yielded the strongest correlation both in horizontal and vertical flame resistance tests. These findings allow us to draw a conclusion that jute fiber and fly ash could be added into PMCs to improve flame resistance properties.

Saiteja et al. [43], in a research study, have gone ahead to determine the mechanical properties of jute-fiber/carbon-based NFRHCs in five different proportions corresponding to 2%, 4%, 6%, 8%, and 10%. The study came up with a resolution that 3 vol% showed enhanced mechanical properties.

Results obtained from the research studies reveal that the mechanical properties of jute-based NFRHC are improved with MWCNTs. Similarly, Singh et al. [44] studied the modal and harmonic behavior of jute and hemp fiber-based NFRHCs. In this chapter, experimental and numerical analyses have been performed on PMCs. In this study, experimental and numerical analyses are conducted on PMCs. It has been documented that hemp-epoxy-based NFRHCs emerged stronger and stiffer compared to jute-epoxy-based NFRHCs. Furthermore, in the given study, more layers of fiber proved to have a significant influence upon increasing the strength and stiffness of the PMCs. It means that hemp fibers could more appropriately give better strength and rigidity to PMCs.

Besides, Senthilrajan et al. [45] reported the investigation on the influence of different fiber parameters on jute-fiber/polyester-based NFRHCs' acoustics for varying fiber lengths such as 10, 15, 20, and 25 mm. A longitudinally oriented fiber with 5 mm and 25% by weight showed a maximum natural frequency owing to its lengthier nature, thereby enhancing the sound absorption capacity. These PMCs consequently represent very attractive potentials for low-cost and high-performance-grade alternatives compared to commercially available damping panels. Therefore, jute fibers can be a promising reinforcement material in PMCs that call for good acoustic and vibrational behavior.

9.2.3 Hemp-fiber-based NFRCs

Hemp plants are derived from the stalk of *Cannabis sativa* and are among the most powerful natural fibers. Naturally, they are resistant to mold and mildew, and have very good insulation properties [38]. It is also highly absorbent with good moisture-wicking properties; therefore, it is highly sought after for making clothes and textiles. Besides, hemp fibers are completely biodegradable and ecological because their growth does not require any harmful chemicals and pesticides [46]. Since the last ten years much attention has been paid to hemp fibers in automotive and aerospace industries as a reinforcing element because of their high strength and stiffness.

Ribeiro et al. [47] addressed the mechanical and thermal properties of hemp-fabric-based NFRHCs. In their work, the addition of hemp fibers up to 30% by volume produced an exceptional enhancement of the developed PMC's mechanical properties. More precisely, an increase in tensile strength, elastic modulus, and Izod impact energy by more than 60%, 80%, and 7.5 times, respectively, was observed. The thermal performance indicated a 20% reduction in the temperature level and indicated that this hemp fabric can potentially be used as reinforcement for epoxy PMCs, especially in those applications that have superior mechanical and ballistic properties. Dayo et al. [48] investigated the influence of different chemical treatments on the mechanical and thermal properties of hemp-fiber/polybenzoxazine-based green composites. The author concluded that silane molecules can interact with the functional groups on the surfaces of fibers as well as with the polymer matrix. Furthermore, the resulting polymer matrix composite exhibits enhanced mechanical and thermal qualities of being robust, lightweight, and eco-friendly.

Likewise, Sahayaraj et al. (2021) [49] examined the impact of chemically treated jute and hemp fibers by analyzing their individual characteristics. The effect of chemical treatment on the mechanical and thermal properties of jute and hemp fibers was analyzed using FTIR and XRD. The findings from the observations revealed that the hybrid composite with 40 wt% jute fibers (JF) and 10 wt% hemp fibers (HF) exhibited a reduced void content, indicating improved compatibility between the matrix and fibers, which is crucial for enhancing mechanical performance. Moreover, the treatment of fibers can lead to enhanced thermal degradation of NFRHCs. The author concluded from the findings obtained that surface treatment enhances the interlocking connections between the fiber and the matrix of the hybrid composite.

Stelea et al. [50] conducted a study where they studied the impact of hemp/matrix ratio and the type of polymer matrix. They specifically focused on analyzing the changes in mechanical properties, thermal stability, and matrix comparison. The outcome from findings demonstrated that the addition of fiber up to 50 wt% enhanced the mechanical properties and reduced the thermal stability. The results indicate that modifying the hemp/matrix ratio and selecting an appropriate matrix type will improve performance characteristics, rendering these composites acceptable for various applications.

9.2.4 Banana-fiber-based NFRCs

Pelissari et al. [51] identified the extraction of cellulose nanofibers (CNF) from banana peel by chemical and mechanical treatments and the preparation of PMCs when mixed in a suspension of banana starch. Addition of CNF to the starch matrix improved the mechanical properties, including increased tensile strength and Young's modulus. This type of PMCs has shown very good thermal stability along with higher water resistance, opacity, and crystallinity. However, excessive mechanical treatment causes defects in the fibers and deteriorates the properties of the PMCs. The five passages in a high-pressure homogenizer were determined to be the optimal condition for the preparation of CNFs and PMCs.

Jandas et al. [52] had aimed at enhancing the processability and economic viability of PLA, which is a biodegradable alternative to petroleum-based plastics, through the incorporation of treated banana fibers and organically derived clays within the grade of cloisite 30 B. In fact, to enhance the interaction with PLA, some of the chemical modifications were carried out on BF; afterward, the mechanical, thermal decomposition, and burning rate features of the optimized composition of the BF-reinforced PLA bio-based PMCs were assessed. Later, bio-based PMCs, under industrial conditions, have been used in the production of biodegradable cutlery.

In their effort to contribute a solution to this world problem, Kamel et al. [53] developed an effective, fast-action wound dressing by incorporating banana peel in chitosan. The peels were dried, pulverized into powder and then incorporated into chitosan at weight percentages of 0%, 2%, 5%, and 10%; the agent for plasticization and crosslinking was glycerol.

While TEM was used to analyze the shape and dimensions of the resultant powder particles, SEM was employed for the analysis of the distribution of the latter within the membranes. The interaction between banana peel powder and chitosan was examined by Fourier Transform Infrared

Spectroscopy. The dielectric loss, dielectric constant, and conductivity were measured in the frequency range from 100 Hz to 100 kHz. Introduction of banana peel powder exhibited several relaxation processes and shifted the loss factor curves to lower frequencies. Antimicrobial activity was highest in membranes with 10% banana peel powder against *Candida albicans*.

9.2.5 Ramie-fiber-based NFRCs

The effect of poly(oxyethylene) (PEO) on high aspect ratio cellulose whiskers in the fabrication of PMC film was investigated by Alloin et al. [54]. The films were fabricated using extrusion as well as casting/evaporation methods. The impact of the material properties ratio, cellulose whiskers' morphology and their distribution, along with factors such as melting temperature, glass transition temperature, thermal stability, and rheological behavior, was observed in the experiments. According to the authors, cast/evaporated PMCs films showed thermal stabilization of modulus at temperatures above the melting point of PEO, while rheological characterization indicates that the PMCs films behave like solid materials, suggesting that cellulose whiskers provide effective reinforcement. The authors concluded that optimizing processing conditions or investigating different reinforcing agents could further enhance the properties of polymer PMCs based on these results.

Syafri et al. [55] carried out the preparation and characterization of composite materials through the strengthening of cellulose fibrils extracted from ramie fiber by the addition of tapioca starch PCC. The CNF-ramie was prepared through a chemical-ultrasonication method, and the composites were produced via the casting solution containing glycerol as a plasticizer. The morphology of the composites was characterized using SEM, FTIR, XRD, and thermogravimetric stability. Correspondingly, CS/4CNF/6PCC had the highest tensile strength, 12.84 MPa. The result for percentage crystallinity obtained was 30.76% for CS/4CNF/6PCC. Thus, from the results obtained, one can deduce that with the inclusion of CNF-ramie and PCC, improvements were found in moisture absorption properties, crystallinity, and thermal stability. SEM images indicated good incorporation of CNF-ramie in the matrix; however, regarding PCC, the interaction with the starch matrix was weaker, probably because of the presence of calcium clumps.

Siqueira et al. [56] produced and characterized whisker-reinforced polycaprolactone composite films in terms of tensile tests and dynamic mechanical analysis, thermal behavior, crystalline behavior, and elemental analysis. Films were derived from ramie and luffa cylindrica using casting raw and chemically modified n-octadecyl isocyanate grafting agents. The applications of modified grafting agents promote their compatibility with the matrix, leading to PMC films with significantly increased mechanical and thermal properties. They observed a notable improvement in the material properties within the context of the environmental problems, specifically wound infections in PMC films. The author determined that whisker-reinforced polycaprolactone PMC films, sourced from ramie and luffa, are promising; however, additional research and optimization are required to advance these novel materials for practical use.

9.3 CRITICAL ANALYSIS OF USAGE OF DIFFERENT MATRIX PHASES

Because of their ease in processing and the formation of strong chemical bonding with bast fibers, polymer matrices find the most common application in bast-fiber-reinforced polymer matric composites. Epoxy, polyester, and polypropylene represent three of the most widely used polymer matrices. Epoxy is often used because it possesses excellent adhesion, chemical resistance, and dimensional stability, while polyester can be employed as an economical polymer with good toughness and stiffness. Being light and cheap, polypropylene suffers from having low modulus and reduced thermal stability, limiting its use in more demanding applications.

The other possible application of ceramic matrices in bast-fiber-reinforced PMCs is seldom realized due to their high processing temperature and consequent cost. Nevertheless, good thermal and wear resistance from the ceramic matrices may also qualify the materials for use at high temperatures and wear. Bast fibers can also be reinforced by other inorganic matrices, such as cement. The PMCs are usually used in the construction of buildings owing to their good strength and durability. Unfortunately, these materials are relatively heavy and brittle; the curing process is also slow and complex.

9.4 INFLUENCE OF VARIOUS TECHNIQUES ON THE BAST-FIBER-BASED NFRHCS

Bast-fiber-based NFRHCs can be manufactured using several techniques, which differently affect the properties of the resulting material. Some of the most common manufacturing techniques are hand layup, resin transfer molding (RTM) [57], compression molding, injection molding, and filament winding, as shown in Figure 9.2. In addition, bast fiber-based NFRHC manufacturing technique is chosen depending on the desired properties of PMCs, production volume, and cost.

9.4.1 Hand layup

The hand layup technique is a widely used method for manufacturing biocomposites on a large scale. It is the simplest and most economical molding technique, applicable to various products. Gel coating is applied to the

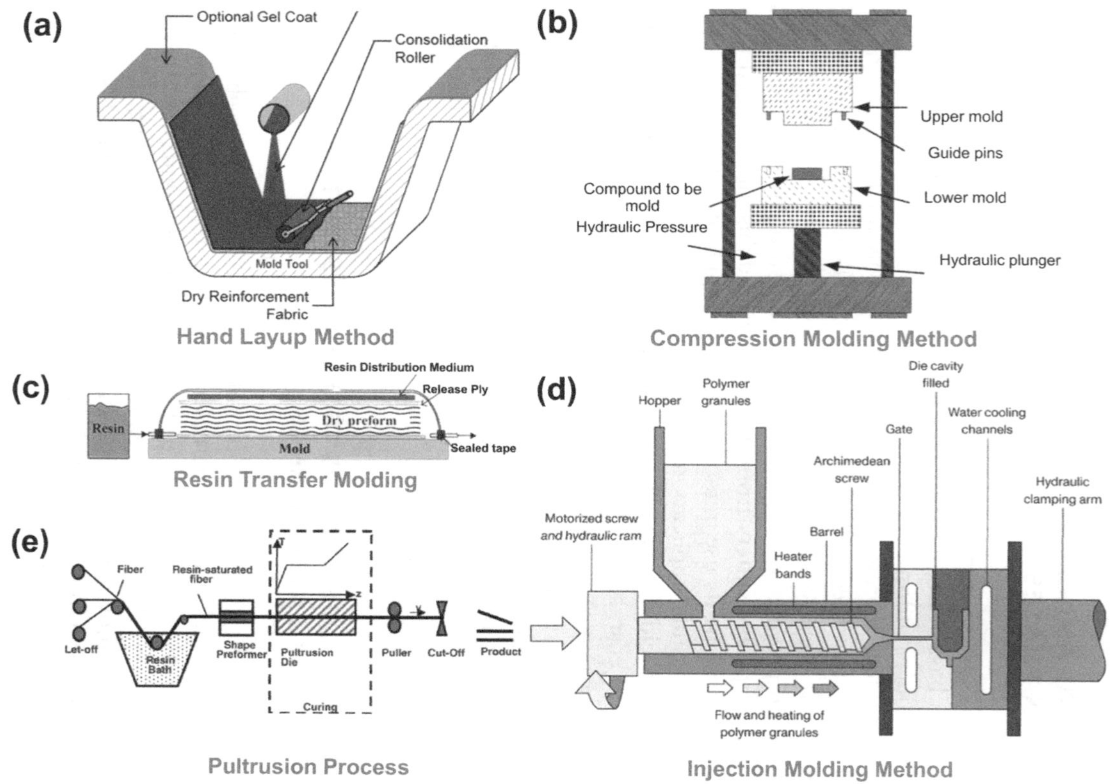

Figure 9.2 Technologies used to manufacture continuous fiber-reinforced composite structures are: (a) hand layup, (b) compression molding, (c) resin Transfer molding, (d) vacuum-assisted resin transfer molding, (e) filament winding [58].

mold surface for easy removal, and natural fiber mats are manually placed inside. A thin resin coat is applied, and a hand roller is used to distribute it evenly. This method is also used for developing particulate composites, where reinforcing long fibers are chopped into short fibers and mixed with resin. After sufficient curing time, the composite is removed for further processing. However, it has limitations such as fiber delamination and time consumption.

9.4.2 Compression molding technique

Compression molding is a large-volume and high-pressure process used to make the composites. It is inexpensive and capable of manufacturing enormous, complicated shapes. Composites can be created utilizing unidirectional matting, bi-directional textiles, randomly oriented composites, or chopped strands. The technique involves depositing charge inside a mold cavity, which is subsequently closed, and pressure is used to push the charge to conform to the mold cavity's shape.

9.4.3 Resin transfer molding technique

This low-pressure and low-temperature technique involves transferring liquid thermoset resin inside a mold made of fiber mats. It is popular for large-scale manufacturing of high-performance structures. A vacuum bag applies pressure, and natural fiber mats and peel ply are placed. Two tubes pump the resin, and extra resin is removed. The resin can be transferred directly or from the side.

9.4.4 Injection molding technique

This process involves feeding raw materials in pellets, granules, or powder into a barrel, which is heated and sheared using a screw. The resin melts and mixes with the fiber pellets, compacting and injecting the molten composite into the mold cavity. The process consists of feed, compacting, and metering zones, with electric motors driving screws. The main advantage is near-net shape with good surface finish and dimensional accuracy.

9.4.5 Pultrusion technique

The pulsed continuous manufacturing process is used to create profiles for composites, using roving fibers pulled through a resin bath. The resin is infused with the fibers, shaped, and cut to the desired length. This process is also used for injection molding pellets. The challenge lies in controlling viscosity and temperature.

9.5 APPLICATIONS OF BAST-FIBER-BASED NFRHCS

Bast hybrid composites, derived from different sections of plants, have been investigated for their superior stability and desired performance, making them ideal for a wide range of applications. These composites can be modified to further decrease moisture uptake and improve compatibility with hydrophobic polymers. Furthermore, the addition of flame retardants enhances their functionality of bast-fiber-based NFRHCs for fire safety in sectors like automotive and aerospace. Bast fibers are lightweight and of superior strength, making them suitable for automotive components such as door panels, armrests, and seat backs. Moreover, their use contributes to reduced vehicle weight, improved fuel efficiency, and lower production costs. PLA hybrid composites are also well-suited for automotive parts, electronics, interior designs, and agricultural products, thanks to their biodegradability and outstanding mechanical characteristics. They outperform wood in several aspects, including dimensional stability, weather resistance, impact resistance, and flame resistance, while also being more cost-effective due to lower maintenance requirements.

On the other hand, in aircraft, radomes are radar-transparent dome-shaped structures that shield radar antennas from aerodynamic loads, weather conditions, and impacts when struck by objects, and they can be made of composite materials based on bast NFRHCS. For instance, Haris et al. [59] emphasized that kenaf works as an efficient reinforcement for a number of polymers utilized in the manufacturing of aviation radomes. The selection of materials was based on their beneficial qualities, spanning from dielectric to mechanical performance. Furthermore, it was highlighted that the material must tolerate adverse weather conditions and demonstrate superior impact resistance. Consequently, mixing kenaf with other fillers can lead to the fabrication of hybrid materials that offer adequate qualities for manufacturing airplane radome covers. Additionally, the high annual yield of bast fibers, together with their noise absorption and mechanical qualities, gives prospects for the creation of interior components in airplanes. The different manufacturing techniques are depicted in Figure 9.3.

The ability of bast-fiber-reinforced composites to absorb moisture often poses challenges in applications involving water for marine composite. To address the issue of these, hybridization of different renewable fiber composites is done. In this, Fiore et al. [60] developed the hybrid composite containing jute and basalt fiber to enhance the strength and durability under salt-fog conditions. The research results indicated that the resulting composite displayed approximately 98% and 90% recovery of its original stiffness and strength after 30 days of exposure to salt-fog spray treatment.

The automotive sector has been one of the active users in the bast-fiber-reinforced PMCs. Several works have also reported their successful applications in interior components such as door panels, dashboard trims, and

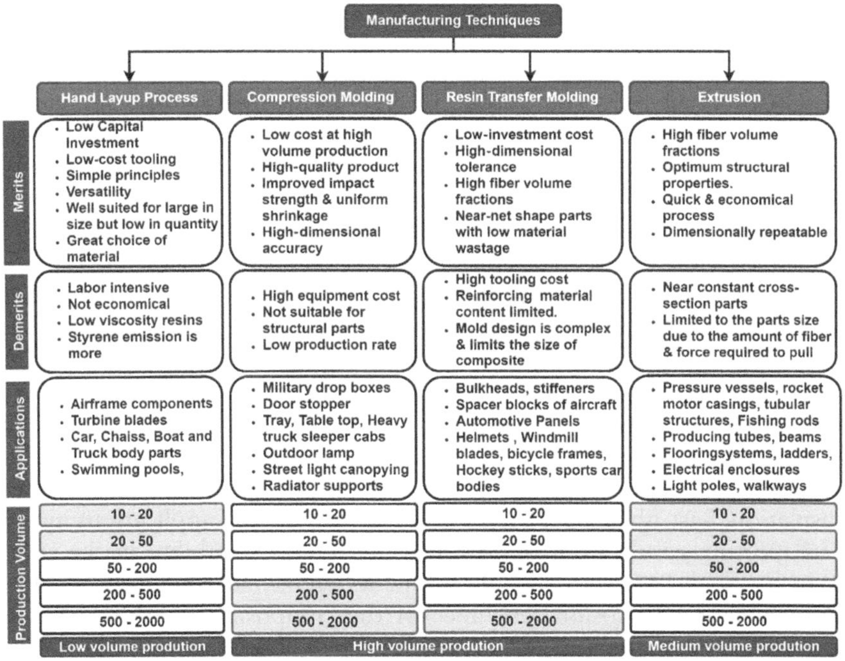

Figure 9.3 Various manufacturing techniques on the bast-fiber-based NFRHCs.

seatbacks. In this context, such PMCs could provide reduced weight, improving fuel efficiency and reducing environmental impacts. Similarly, the aerospace industry recognizes huge potential for bast-fiber-based PMCs in reducing weight and increasing sustainability. It has also found its applications in aircraft interiors: cabin panels, overhead compartments, and flooring are built with these materials. These use the lightweight properties of bast-fiber-based PMCs to provide improved fuel efficiency and lower emissions [61].

Sectoral blasting of fiber-reinforced PMCs has shown great promise for applications in the construction industry. They were researched for manufacturing various environment-friendly building materials like roofing sheets, wall panels, and insulation boards. These PMCs have relative advantages such as thermal and sound insulation, sustainability, and low environmental impact. The sports industry has, on their part, come to appreciate the role of bast-fiber-based PMCs in the performance and durability of sporting goods. Applications include tennis rackets, bicycles, hockey sticks, and skis, among others. Some of the advantages these PMCs offer are high strength, stiffness, and impact resistance, which would enhance athletic performance. It has also shown potential for application in marine use due to its water, moisture, and salt resistance. Applications in boat hulls, decks, and interior components have been explored. In general, these PMCs exhibited superior mechanical properties and corrosion resistance, hence becoming viable options to the conventional materials being used in marine environments. Bast-fiber-based PMCs also hold potential applications in the agricultural sector in plant pots, seedling trays, and irrigation pipes. Their biodegradability and moisture retention are highlighted in some research papers as key attributes to sustainable agriculture.

From various pieces of previous work [1–11, 15, 19, 21, 28], it has been observed that bast-fiber-reinforced polymer composites are used in different industries. Unique properties of the PMCs, in terms of their lightweight nature, high strength, thermal and sound insulation, biodegradability, and sustainability, make them appealing options among conventional materials. While bast-fiber-based PMCs contribute to the reduction in weight for automotive and aerospace industries, improved fuel efficiency and reduced emissions are the after-effects. These PMCs are considered eco-friendly building materials in the construction industry with properties for thermal and sound insulation. Bast-fiber-based PMCs also find widespread applications in various sporting goods to improve performance and durability in the sports industry. Such PMCs are also useful for different marine industries due to the water and corrosion resistance of the composites. In the packaging industry, biodegradability, and compostability bring a low environmental impact [62]. Natural looks and sustainability attract consumer goods such as furniture and electronic casings. In the case of agriculture, it can replace plant pots, seedling trays, and irrigation pipes with sustainable materials. Therefore, the reviewed literature emphasized wide-ranging applications of bast-fiber-reinforced PMCs in diverse industries. Further studies and development of these PMCs would offer good opportunities for overcoming the problems in sustainability and actually contributing to a circular economy and Sustainable Development Goals (SDG).

9.6 CONCLUSION

The use of bast fibers has gained significant interest due to their potential in various industries, including textiles, building construction, automotive manufacturing, aerospace, and automotive manufacturing. Bast fibers offer excellent physicomechanical properties, high strength, and durability, making them suitable for producing durable textiles and products such as clothing and industrial fabrics. They also create functionalized composite materials for building construction, exhibiting good strength, fire resistance, and unique characteristics. These bast NFRHCs are created using different polymeric substances and processing methods, with fibers undergoing chemical modifications to enhance adhesion toward sustainable and cleaner production. Moreover, biodegradable PMC films generated from bast fibers are eco-friendly alternatives to traditional synthetic plastics. Conversely, incorporation of fillers into bast NFRHCs would enhance fiber surface properties, leading to increased functionalities like antibacterial, waterproof, and UV-resistant properties. However, scaling and commercializing bast fiber composites raises concerns about production costs and sustainability.

REFERENCES

1. Summerscales, J., 2021. A review of bast fibres and their composites: Part 4 ~ organisms and enzyme processes. *Composites Part A: Applied Science and Manufacturing*, 140, 106149.
2. Saleem, A., Medina, L., Skrifvars, M., and Berglin, L., 2020. Hybrid polymer composites of bio-based bast fibers with glass, carbon and basalt fibers for automotive applications: A review. *Molecules 2020*, 25 (21), 4933.
3. Kiruthika, A. V., 2017. A review on physico-mechanical properties of bast fibre reinforced polymer composites. *Journal of Building Engineering*, 9, 91–99
4. Mohanty, A.K., Misra, M., and Drzal, L.T., eds., 2005. *Natural Fibers, Biopolymers, and Biocomposites*. Routledge, 896.
5. Divya, D., Devi, S.Y., Indran, S., Raja, S., and Sumesh, K.R., 2022. Extraction and modification of natural plant fibers: A comprehensive review. *Plant Fibers, Their Composites, and Applications*, 2022, 25–50.
6. Rao, K.M.M. and Rao, K.M., 2007. Extraction and tensile properties of natural fibers: Vakka, date and bamboo. *Composite Structures*, 77 (3), 288–295.
7. Moujoud, Z., Sair, S., Ait Ousaleh, H., Ayouch, I., El Bouari, A., and Tanane, O., 2023. Geopolymer composites reinforced with natural Fibers: A review of recent advances in processing and properties. *Construction and Building Materials*, 388, 131666.
8. Lotfi, A., Li, H., Dao, D.V., and Prusty, G., 2019. Natural fiber–reinforced composites: A review on material, manufacturing, and machinability. *Journal of Thermoplastic Composite Materials*, 34 (2), 238–284. https://doi.org/10.1177/0892705719844546
9. Mallick, P.K., 2007. *Fiber-Reinforced Composites: Materials, Manufacturing, and Design*, Third Edition. Taylor & Francis.
10. Mitschang, P., 2011. Manufacturing of thermoplastic fiber-reinforced polymer composites. *Wiley Encyclopedia of Composites*, 2011, 1–15.
11. Ho, M.P., Wang, H., Lee, J.H., Ho, C.K., Lau, K.T., Leng, J., and Hui, D., 2012. Critical factors on manufacturing processes of natural fibre composites. *Composites Part B: Engineering*, 43 (8), 3549–3562.
12. Park, C.H. and Lee, W.I., 2012. Compression molding in polymer matrix composites. *Manufacturing Techniques for Polymer Matrix Composites (PMCs)*, 2012, 47–94
13. Oksman, K., 2016. High quality flax fibre composites manufactured by the resin transfer moulding process. 20 (7), 621–627. https://doi.org/10.1177/073168401772678634
14. Chattopadhyay, S.K., Khandal, R.K., Uppaluri, R., and Ghoshal, A.K., 2011. Bamboo fiber reinforced polypropylene composites and their mechanical, thermal, and morphological properties. *Journal of Applied Polymer Science*, 119 (3), 1619–1626.

15. Neuba, L. de M., Pereira Junio, R.F., Ribeiro, M.P., Souza, A.T., Lima, E. de S., Filho, F. da C.G., Figueiredo, A.B.H. da S., Braga, F. de O., de Azevedo, A.R.G., and Monteiro, S.N., 2020. Promising mechanical, thermal, and ballistic properties of novel epoxy composites reinforced with *Cyperus malaccensis* sedge fiber. *Polymers 2020*, 12 (8), 1776.
16. Djafari Petroudy, S.R., 2017. Physical and mechanical properties of natural fibers. In: *Advanced High Strength Natural Fibre Composites in Construction*. Elsevier, 59–83.
17. Ramanaiah, K., Ratna Prasad, A. V., and Hema Chandra Reddy, K., 2011. Thermal and Mechanical Properties of Sansevieria Green Fiber Reinforcement, 16 (8), 602–608. https://doi.org/10.1080/1023666X.2011.622358
18. Mir, S.S., Hasan, S.M.N., Hossain, M.J., and Hasan, M., 2012. Chemical modification effect on the mechanical properties of coir fiber. *Engineering Journal*, 16 (2), 73–84.
19. El-Sakhawy, M. and Hassan, M.L., 2007. Physical and mechanical properties of microcrystalline cellulose prepared from agricultural residues. *Carbohydrate Polymers*, 67 (1), 1–10.
20. Komuraiah, A., Kumar, N.S., and Prasad, B.D., 2014. Chemical composition of natural fibers and its influence on their mechanical properties. *Mechanics of Composite Materials*, 50 (3), 359–376
21. Suyambulingam, I., Raja, S., Divakaran, D., Sanjay, M.R., Siengchin, S., and Santhi, R., 2022. Green methods for surface modification of bast fibers. In: Rajeshkumar, G., Devnani, G., Sinha, S., Sanjay, M., Siengchin, S. (eds.), *Bast Fibers and Their Composites. Springer Series on Polymer and Composite Materials*. Springer. https://doi.org/10.1007/978-981-19-4866-4_5
22. Rani, M.S.A., Rudhziah, S., Ahmad, A., and Mohamed, N.S., 2014. Biopolymer electrolyte based on derivatives of cellulose from kenaf bast fiber. *Polymers*, 6 (9), 2371–2385.
23. Kargarzadeh, H., Ahmad, I., Abdullah, I., Dufresne, A., Zainudin, S.Y., and Sheltami, R.M., 2012. Effects of hydrolysis conditions on the morphology, crystallinity, and thermal stability of cellulose nanocrystals extracted from kenaf bast fibers. *Cellulose*, 19 (3), 855–866.
24. Ntenga, R., Saïdjo, S., Beda, T., and Béakou, A., 2019. Estimation of the effects of the cross-head speed and temperature on the mechanical strength of kenaf bast fibers using Weibull and Monte-Carlo statistics. *Fibers*, 7 (10), 1–9.
25. Roy, S. and Lutfar, L.B., 2012. Bast fibres: Jute. *Handbook of Natural Fibres: Second Edition*, Woodhead Publishing Series in Textiles. Elsevier, 39–59.
26. Assarar, M., Scida, D., El Mahi, A., Poilâne, C., and Ayad, R., 2011. Influence of water ageing on mechanical properties and damage events of two reinforced composite materials: Flax–fibres and glass–fibres. *Materials & Design*, 32 (2), 788–795.
27. Amroune, S., Belaadi, A., Bourchak, M., Makhlouf, A., and Satha, H., 2020. Statistical and experimental analysis of the mechanical properties of flax fibers, 19 (4), 1387–1401. https://doi.org/10.1080/15440478.2020.1775751
28. Yan, L. and Chouw, N., 2013. Crashworthiness characteristics of flax fibre reinforced epoxy tubes for energy absorption application. *Materials & Design*, 51, 629–640.
29. Sathish, S., Kumaresan, K., Prabhu, L., Gokulkumar, S., Karthi, N., and Vigneshkumar, N., 2019. Experimental investigation of mechanical and

morphological properties of flax fiber reinforced epoxy composites incorporating SiC and Al_2O_3. *Materials Today: Proceedings*, 2249–2253. https://doi.org/10.1016/j.matpr.2019.09.106

30. Santulli, C., Janssen, M., and Jeronimidis, G., 2005. Partial replacement of E-glass fibers with flax fibers in composites and effect on falling weight impact performance. *Journal of Materials Science*, 40, 3581–3585. https://doi.org/10.1007/s10853-005-2882-y
31. Li, X., Tabil, L.G., Panigrahi, S., and Crerar, W.J., n.d. *Processing Flax Fiber-reinforced Polymer Composites by Injection Molding Written for presentation at the CSAE/SCGR 2005 Meeting Winnipeg*, Manitoba.
32. Giuliani, P.M., Giannini, O., and Panciroli, R., 2023. Creep and stress relaxation of unidirectional flax fiber reinforced laminates. *Composite Structures*, 310, 116755.
33. Rajesh, D., Lenin, N., Cep, R., Anand, P., and Elangovan, M., 2023. Enhancement of thermal behaviour of flax with a ramie fibre-reinforced polymer composite. *Polymers 2023*, 15 (2), 350.
34. Paulo, A., Santos, J., da Rocha, J., Lima, R., Ribeiro, J., 2023. Mechanical properties of PLA specimens obtained by additive manufacturing process reinforced with flax fibers, *Journal of Composites Science*, 7, 27. https://doi.org/10.3390/jcs7010027
35. Zhu, J., Zhu, H., Njuguna, J., and Abhyankar, H., 2013. Recent development of flax fibres and their reinforced composites based on different polymeric matrices. *Materials 2013*, 6 (11), 5171–5198.
36. Samant, L., Goel, A., Mathew, J., Jose, S., and Thomas, S., 2023. Effect of surface treatment on flax fiber reinforced natural rubber green composite. *Journal of Applied Polymer Science*, 140 (12), e53651.
37. Sever, K., Sarikanat, M., Seki, Y., Erkan, G., Erdoĝan, Ü.H., and Erden, S., 2012. Surface treatments of jute fabric: The influence of surface characteristics on jute fabrics and mechanical properties of jute/polyester composites. *Industrial Crops and Products*, 35 (1), 22–30.
38. Ahmad, A., Moeed, A., Tariq, A., Nawab, Y., Farooq, M.U., Umair, M. 2022. Effect of hybridization approach on mechanical performance of jute-hemp/epoxy hybrid composite laminates. https://doi.org/10.1080/15440478.2022.2064398
39. Prabhu, L., Krishnaraj, V., Sathish, S., GokulKumar, S., and Karthi, N., 2020. Study of mechanical and morphological properties of jute-tea leaf fiber reinforced hybrid composites: Effect of glass fiber hybridization. *Materials Today: Proceedings*, 27 (3), 2372–2375.
40. Rezghi Maleki, H., Hamedi, M., Kubouchi, M., and Arao, Y., 2018. Experimental study on drilling of jute fiber reinforced polymer composites, 53 (3), 283–295. https://doi.org/10.1177/0021998318782376
41. Sajin, J.B., Babu Aurtherson, P., Binoj, J.S., Manikandan, N., Senthil Saravanan, M.S., and Haarison, T.M., 2021. Influence of fiber length on mechanical properties and microstructural analysis of jute fiber reinforced polymer composites. *Materials Today: Proceedings*, 39, 398–402.
42. Balan, G.S., Balasundaram, R., Chellamuthu, K., Gopan, S.N., Dinesh, S., Vijayan, V., Sathish, T., and Rajkumar, S., 2022. Flame resistance characteristics of woven jute fiber reinforced fly ash filled polymer composite. *Journal of Nanomaterials*, 2022 (5), 1–12.

43. Saiteja, J., Jayakumar, V., and Bharathiraja, G., 2020. Evaluation of mechanical properties of jute fiber/carbon nano tube filler reinforced hybrid polymer composite. *Materials Today: Proceedings*, 22, 756–758.
44. Singh, S.P., Dutt, A., and Hirwani, C.K., 2022. Mechanical, modal and harmonic behavior analysis of jute and hemp fiber reinforced polymer composite. *Journal of Natural Fibers*, 20 (1), 1–15. https://doi.org/10.1080/15440478.2022.2140328
45. Senthilrajan, S., Venkateshwaran, N., and Naresh, K., 2022. Parametric study of different fiber parameters and their influence on acoustics and vibration behavior of jute fiber/polyester resin composites, 19(16), 13063–13075. https://doi.org/10.1080/15440478.2022.2085225
46. Kanimozhi, T., Sundara Pandian, S., Gokulkumar, S., and Sathish, S., 2022. Influence of *Acacia concinna* and *Vachellia nilotica* seed nanopowder on the properties of short Turkish hemp–reinforced epoxy composites. *Proceedings of the Institution of Mechanical Engineers, Part L: Journal of Materials: Design and Applications*, 146442072211454.
47. Ribeiro, M.P., Neuba, L.D.M., Da Silveira, P.H.P.M., Da Luz, F.S., Figueiredo, A.B.H.D.S., Monteiro, S.N., and Moreira, M.O., 2021. Mechanical, thermal and ballistic performance of epoxy composites reinforced with Cannabis sativa hemp fabric. *Journal of Materials Research and Technology*, 12, 221–233.
48. Dayo, A.Q., Gao, B.C., Wang, J., Liu, W.B., Derradji, M., Shah, A.H., and Babar, A.A., 2017. Natural hemp fiber reinforced polybenzoxazine composites: Curing behavior, mechanical and thermal properties. *Composites Science and Technology*, 144, 114–124.
49. Sahayaraj, A.F., Muthukrishnan, M., Ramesh, M., and Rajeshkumar, L., 2021. Effect of hybridization on properties of tamarind (Tamarindus indica L.) seed nano-powder incorporated jute-hemp fibers reinforced epoxy composites. *Polymer Composites*, 42 (12), 6611–6620.
50. Stelea, L., Filip, I., Lisa, G., Ichim, M., Drobotă, M., Sava, C., and Mureşan, A., 2022. Characterisation of hemp fibres reinforced composites using thermoplastic polymers as matrices. *Polymers 2022*, 14 (3), 481.
51. Pelissari, F.M., Andrade-Mahecha, M.M., do Amaral Sobral, P.J., and Menegalli, F.C., 2017. Nanocomposites based on banana starch reinforced with cellulose nanofibers isolated from banana peels. *Journal of Colloid and Interface Science*, 505, 154–167.
52. Jandas, P.J., Mohanty, S., and Nayak, S.K., 2013. Surface treated banana fiber reinforced poly (lactic acid) nanocomposites for disposable applications. *Journal of Cleaner Production*, 52, 392–401.
53. Kamel, N.A., Abd El-Messieh, S.L., and Saleh, N.M., 2017. Chitosan/banana peel powder nanocomposites for wound dressing application: Preparation and characterization. *Materials Science and Engineering: C*, 72, 543–550.
54. Alloin, F., D'Aprea, A., Dufresne, A., Kissi, N. El, and Bossard, F., 2011. Poly(oxyethylene) and ramie whiskers based nanocomposites: Influence of processing: Extrusion and casting/evaporation. *Cellulose*, 18 (4), 957–973.
55. Syafri, E., Kasim, A., Abral, H., Sulungbudi, G.T., Sanjay, M.R., and Sari, N.H., 2018. Synthesis and characterization of cellulose nanofibers (CNF) ramie reinforced cassava starch hybrid composites. *International Journal of Biological* Macromolecules, 120, 578–586.

56. Siqueira, G., De Menezes Jr., A., Bras, J., and Dufresne, A., n.d. *Proceedings of the International Convention of Society of Wood Science and Technology and United Nations Economic Commission for Europe – Timber Committee*, October 11–14, 2010, Geneva, Switzerland.
57. Kang, M.K., Lee, W.I., and Hahn, H.T., 2000. Formation of microvoids during resin-transfer molding process. *Composites Science and Technology*, 60 (12–13), 2427–2434.
58. Santhosh Kumar, S., and Hiremath, S. S., 2020. Natural fiber reinforced composites in the context of biodegradability: A review. In: *Encyclopedia of Renewable and Sustainable Materials*. Elsevier, 160–178. https://doi.org/10.1016/B978-0-12-803581-8.11418-3
59. Haris, M.Y., Laila, D., Zainudin, E.S., Mustapha, F., Zahari, R., Halim, Z., 2011. Preliminary review of biocomposites materials for aircraft radome application. *Key Engineering Materials*, 471–472, 563–567. https://doi.org/10.4028/www.scientific.net/KEM.471-472.563
60. Fiore, V., Scalici, T., Sarasini, F., Tirilló, J., Calabrese, L., 2017. Calabrese, salt-fog spray aging of jute-basalt reinforced hybrid structures: Flexural and low velocity impact response. *Composites Part B: Engineering*, 116, 99–112. https://doi.org/10.1016/j.compositesb.2017.01.031
61. Mathew, L., Joseph, K.U., and Joseph, R., 2006. Swelling behaviour of isora/natural rubber composites in oils used in automobiles. *Bulletin of Materials Science*, 29 (1), 91–99.
62. Zhang, X., Tanguy, N.R., Chen, H., Zhao, Y., Gnanasekar, P., Le Lagadec, R., and Yan, N., 2022. Lignocellulosic nanofibrils as multifunctional component for high-performance packaging applications. *Materials Today Communications*, 31, 103630.

Chapter 10

Advances in self-healing composites

Classification, mechanisms, applications, and future perspectives

Rituparna Duarah, Jayaramudu Jarugala, and Emmanuel Rotimi Sadiku

10.1 INTRODUCTION

In recent years, notable advancements in polymer material development have spurred rapid industrial expansion, leading to a significant surge in plastic manufacturing and consumption [1]. This surge has led to a considerable increase in plastic waste and has raised significant environmental issues [1, 2]. A key factor contributing to polymer disposal is their tendency to lose their original appearance and functionality over time. To address this issue, the adoption of self-healing systems that can autonomously repair damage induced by external factors and thereby extend the lifespan of materials has emerged as an effective strategy for reducing polymer waste [3, 4]. Self-healing techniques involve the adjustment of molecular structures to enable auto-repair of broken or damaged parts in dynamic systems, thereby extending their lifespan. The principle of self-healing is inspired by the innate regenerative abilities found in living organisms, which have the capacity to recover from damage [5]. Over the last decade, significant research efforts have focused on developing self-healing properties in materials, leading to notable advancements in this area [6, 7]. The development of self-healing composites marks an important milestone in material science, aiming to enhance the resilience and longevity of materials by enabling them to autonomously repair damage. This capability promises transformative applications across various industries. Similar to living organisms, self-healing materials are engineered to recover their original properties and characteristics after being damaged. These systems provide protection to outer layers, shielding the delicate core of metallic materials from performance degradation due to physical harm. They also enhance their intrinsic features, contributing to both self-repair abilities and overall device functionality [5, 6]. As a result, self-healing materials are widely employed as protective coatings for electronic devices and vehicles, effectively extending their lifespan. This is accomplished by reducing damage to external surfaces, preserving internal circuitry, and preventing the formation of cracks or scratches. Over time, numerous researchers have introduced new approaches to categorize self-healing into intrinsic and extrinsic processes.

 DOI: 10.1201/9781003527275-10

Recent advancements have shown that self-healing materials can achieve remarkable electrochemical or thermal performance [6]. However, a major challenge persists: balancing their properties and features to enable widespread use. To overcome this, developing innovative molecular designs for self-healing materials is essential, not only to reduce plastic waste but also to increase the durability of materials. Self-healing materials imitate the natural repair mechanisms found in living organisms, which recover from damage to maintain their functionality. These composites have emerged from the convergence of advanced material design and engineering principles, incorporating innovative mechanisms that enable them to detect and respond to damage autonomously. The potential benefits are vast, including extended service life, enhanced safety, and reduced maintenance costs, offering a new paradigm for the future of durable and resilient materials [4–7].

This chapter provides a summary of the self-healing techniques, present-day state of self-healing composites, exploring the underlying principles, mechanisms, and future perspectives on their development. Through this exploration, we aim to illuminate the transformative potential of self-healing composites and their role in the future of materials science and engineering.

10.2 INTRINSIC SELF-HEALING

Intrinsic self-healing materials exhibit a remarkable capacity to repair micro- or molecular-scale damage by leveraging the natural reversibility of bonds within the polymer matrix. This capability typically stems from the unique molecular structures of these polymers, which enable significant inter-chain mobility when subjected to modest energy inputs such as heat, static load, or UV radiation [8, 9]. Upon removal of the stimulus, the material undergoes a process that restores the physical or chemical bond strength. A key advantage of intrinsic healing over extrinsic methods is the ability to fully or partially restore the material's original properties multiple times. Additionally, this intrinsic healing method is applicable to thermoplastics, thermosets, and elastomers [10].

10.2.1 Reversible covalent bonds

When damage occurs, the dynamic formation and breaking of covalent bonds are critical for enabling stress-bearing bonds to dissociate and reform, to facilitate the rapid structural realignments necessary for the healing process. The effectiveness of this self-healing mechanism depends on the nature of the bonding interactions, molecular flexibility, environmental factors, and the material's architecture. Modern dynamic covalent systems utilize a variety of established reaction mechanisms, including chain-exchange reactions and ring-chain equilibrium in cross-linked poly(dimethylsiloxane) [11], trithiocarbonate exchange reactions [12], disulfide [13], and retro Diels-Alder (DA) reactions [14], as well as reversible hydrazone linkages [14] and alkoxyamines [15, 16]. These mechanisms, along with diarylbenzofuranone [17], are typically triggered by heat, though other stimuli such as irradiation, pH shifts, or catalytic additives have also been reported [14, 18]. In DA reactions, several dienophiles and dienes can be used, with the furan–maleimide interaction being particularly notable, functioning within a 100°C–150°C range of healing temperature [19]. Disulfide bonds have garnered increasing attention due to their versatility and ease of integration into existing networks. These bonds have demonstrated their ability to promote healing in a range of materials: in hybrid sol–gel coatings, in room temperature elastomers *via* aromatic disulfide metathesis, in conventional epoxy-based thermosets with thiols and aliphatic disulfide chains at mild temperatures (60°C–70°C) [20, 21]. Despite the long-known reversible chemistry of disulfide bonds and DA/rDA reactions, their application in creating self-healing polymers has only recently gained traction [21].

10.2.2 Exchange reactions

Dynamic exchange reactions, a form of reversible covalent bonding, offer an effective strategy for developing self-healing polymers. These reactions are particularly beneficial because they require lower activation temperatures compared to DA reactions to initiate the healing process [9]. For instance, disulfide bonds can undergo exchange reactions between adjacent S–S bonds via ionic intermediates or free radicals, enabling the repair of damage or cracks through heating, oxidation, or photolysis at low temperatures [9]. Amamoto et al. reported the incorporation of thiuram disulfide units into polyurethane with a low glass transition temperature (T_g), achieving successful self-healing under visible light at room temperature [22]. Tsarevsky and Matyjaszewski reported the demonstration of strong healing capabilities through the oxidation of thiols from thiol–disulfide exchange reactions in a polystyrene copolymer, which regenerated disulfide bonds [23]. Earlier studies have shown that disulfide bonds can engage in metathesis exchange reactions, where neighboring S–S bonds are simultaneously reformed and cleaved through ionic intermediates or free radicals. Furthermore, the unique features of reversible rearrangement reactions involving S–H and S–S bonds can be harnessed by introducing tri-thiocarbonates. Two thiol (S–H) groups can be formed by reduction of S–S bonds due to its reversible nature, which can be oxidized to re-establish the disulfide (S–S) linkages [24]. This phenomenon holds significant promise for self-healing material applications. Higaki et al. reported that the healing of cracks was triggered by the formation of free radicals upon cracking by incorporating a thermoreversible alkoxyamine reaction into a cross-linked reversible network [25]. Kamada et al. used a modified core-first approach through atom transfer radical polymerization for the synthesis of a star polymer based on poly(n-butyl acrylate), specifically polyEGDA-(polyBA)n [26]. The effectiveness of self-healing in free-radical-based materials depends on the mobility of these radicals and the avoidance of interactions with oxygen. The reactive chain ends must meet and complete the necessary reactions for effective self-healing, prior to any unwanted interactions, such as radical-oxygen reactions or the capture of free radicals. If the free radicals react with oxygen rather than the fractured polymer ends, the healing process will be compromised [27].

10.2.3 Supramolecular chemistry dynamics

Supramolecular interactions are crucial in enabling self-organization and self-assembly through reversible and directional non-covalent forces, in the determination of mechanical attributes of materials [28, 29]. Supramolecular interactions, unlike covalent bonds, enable quick adjustments within the material's structure, allowing it to shift from a fluid-like state with high free volume and low density to a more solid state with reduced free volume and enhanced elasticity or plasticity [30]. However, this type of chemistry does come with drawbacks. These innovations hold the potential to unlock numerous new technological possibilities. The wide range of network structures that are enabled due to supramolecular interactions is highly appealing for applications in biomedical technologies, soft electronics, sensing, 3D/4D printing, and reprocessable materials, offering significant technological potential. Mechanical integrity in these systems is achieved through multiple non-covalent interactions between associative groups attached to polymer chains, including hydrogen bonding, host-guest interactions, metal–ligand complexation, π–π stacking, ionic interactions, and hydrophobic interactions, as shown in Figure 10.1 [30].

Polymers based on supramolecular interactions generally have lower T_g and weaker mechanical strength, which limits their use to certain specialized applications. These reversible interactions are especially suited for the development of intrinsic self-healing polymers [31]. Some of the most researched include the interactions found in ionomers, commonly used in ballistic and coating applications, and hydrogen bonding, as seen in ureidopyrimidinone units and self-complementary hydrogen bonding groups in randomly branched oligomers [32]. Ionomers are polymers where acid groups are neutralized to form ionic metal salts attached to the polymer backbone, resulting in electrostatic interactions [33]. These ionic clusters improve the physical and mechanical properties of the material and are present within the polymer. The self-healing ability of these systems is largely due to the transitions and relaxations of these ionic clusters within specific temperature ranges. Recent research has highlighted the importance of non-neutralized acid groups in facilitating healing at lower temperatures [34]. Sijbesma et al. [35] developed a self-healing elastomer by introducing a robust, reversible polymer system based on quadruple hydrogen bonding through a collaboration between ESPCI and Arkema [36]. It was the first

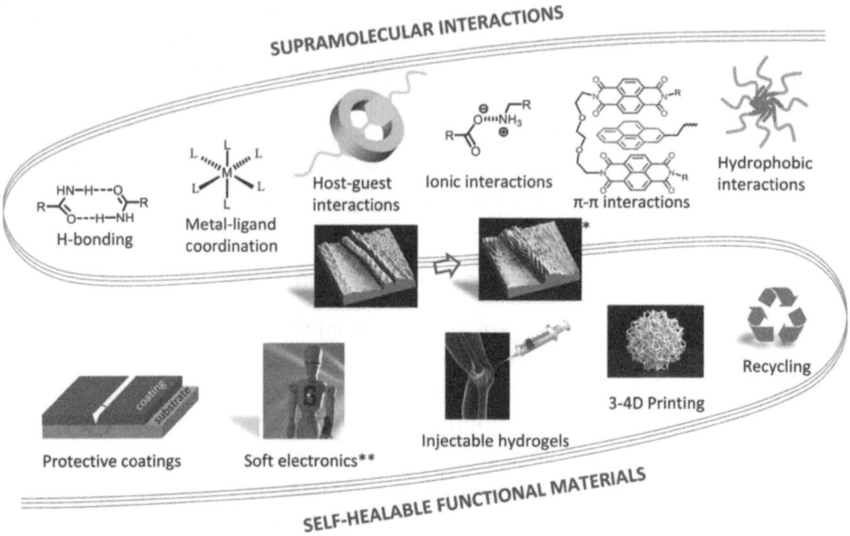

Figure 10.1 Types of supramolecular chemistries employed in self-healing functional materials.

Reproduced from Ref [30] with permission from Wiley Advanced, 2018; Self-healing AFM images: Reproduced from Ref [28] with permission from the Royal Society of Chemistry, 2011; Soft electronics image: Reproduced from Ref [31] with permission from Springer Nature, 2012.

self-healing elastomer available for commercial use. Other supramolecular chemistries, though less explored, include metal coordination and π–π stacking interactions, such as those between pyrene and polydiimide groups [29]. Despite the variety of chemistries supporting intrinsic self-healing, there remains a lack of definitive studies demonstrating that interfacial healing can occur solely through chemical interactions without prior molecular interdiffusion [29, 37]. The physical process of chain interdiffusion is very important to achieve the complete interfacial healing unless the polymer network comprises entirely reversible bonds that can regenerate the same bonds as the original material. This suggests that a universal healing mechanism, irrespective of the specific chemistry, may be explained by the physical model proposed by Wool and O'Connor [38], which includes approach, wetting, surface rearrangement, randomization, chain interdiffusion, and chemical interactions. This signifies the complex interplay between polymer structure and self-healing mechanisms [38].

10.2.4 Hydrogen bonding

Hydrogen bonds play a crucial role in forming a wide array of supramolecular polymers, each with distinct mechanical properties. Although typically weaker than covalent bonds, the directional nature and strong affinity of H-bonds allow them to create robust, rubber-like materials [9]. Research has shown that the strategic integration of multiple hydrogen bonds into functional units, such as urea isopyrimidinone (Upy), can significantly increase the bonding strength between Upy units and various polymers, including polyesters, polyethers, and polysiloxane [39, 40]. Yamauchi et al. demonstrated that incorporating Upy into flexible polymers like poly(ethylene-co-butylene) and polydimethylsiloxane (PDMS) enhances their mechanical strength [41]. Similarly, Meijer and his team utilized reversible hydrogen bonding to create stable DDAA (donor–donor–acceptor–acceptor) dimers ureidopyrimidinone (Upy), a technique that shows promise for developing carbon-based nanocomposites with structural applications [42–44]. Beyond contributing to mechanical strength, hydrogen bonds also facilitate the creation of materials capable of undergoing multiple healing cycles at the same damaged site [43]. This self-healing capability is particularly valuable in extending the material's lifespan and maintaining its performance. However, for the healing process to be effective, the damaged ends of the material must come into contact immediately after breakage. If there is a delay, hydrogen bonds might form with nearby units instead of the broken ends, diminishing the material's ability to heal efficiently [8]. This highlights the importance of timely interaction between the damaged surfaces to ensure the full benefits of hydrogen-bonded supramolecular systems are realized.

10.2.5 Metal–ligand coordination

Metal–ligand coordination represents a unique type of non-covalent supramolecular interaction that takes place between a metal ion, serving as the coordination center, and ligands (surrounding organic molecules) [45]. These coordination bonds offer significant potential for developing self-healing polymers because they form spontaneously and have highly adjustable bond energies. The strength of these bonds lies between that of covalent bonds and van der Waals interactions; however, their primary advantage is the capacity to adjust their strength through the careful optimization of the suitable combination of metal ion and ligand [46]. This adjustability introduces a level of reversibility similar to that found in hydrogen bonds, which is a key feature for materials designed to self-heal. However, the strong nature of metal–ligand coordination bonds can present challenges. The healing process often requires relatively high temperatures to be effective, which may limit the softness and flexibility of the resulting material [47]. Practical considerations must be taken into account to create self-healing materials using any type of ligand by modifying bond strength through the selection of different metal ions. These factors encompass the intricacy of synthesizing the ligand, the accessibility of starting materials, and the difficulties associated with integrating the ligand into a polymer matrix [45]. Moreover, some ligands are water-soluble, which may render them unsuitable for specific applications where water resistance is crucial. Despite these challenges, several ligands have proven successful in metal–ligand coordination for self-healing polymers [46, 47]. Notable examples include catechol, which is known for its strong binding affinity with metal ions, histidine, which offers a balance of coordination strength and flexibility, pyridinyl, and carboxylate groups, which are both versatile and effective in forming stable, reversible bonds [48]. Consequently, the meticulous choice and design of these ligands can help in the development of sophisticated self-healing materials with customized characteristics, positioning metal–ligand coordination as a highly promising approach in the realm of smart polymers.

10.2.6 π–π stacking interactions

π–π stacking interactions, a form of non-covalent bonding, occur between π electron-deficient groups and π electron-rich aromatic molecules. These interactions can be harnessed to develop self-healing materials that respond to thermal stimuli, providing a reversible repair mechanism [49]. A prominent example is the research by Burattini and colleagues, who used π–π stacking between electron-rich pyrenyl units and electron-deficient diimide groups to create a self-healing supramolecular polymer blend [50]. This blend consisted of a telechelic polyurethane with pyrenyl end groups and a chain-folding polyimide, resulting in an elastomeric material with significant self-repair capabilities. Their study revealed that the material could recover up to 95% of its tensile modulus, 977% of its toughness, and 1% of its elongation at break after damage. π–π stacking interactions in self-healing materials has the ability to activate the healing process over a broad temperature range, typically 50°C–100°C. This flexibility is achieved by adjusting the polymer's T_g through modifications to spacer molecules and blend composition. For example, a polymer blend with a linear polysiloxane backbone and folding "tweezer-type" polyimide can heal at 100°C, whereas a polyimide–polyamide network can self-repair at 50°C [29]. This adaptability makes π–π stacking interactions highly versatile to develop self-healing materials for diverse applications. However, the relatively weak bond strength associated with π–π interactions is a limitation of this mechanism. Although sufficient for self-healing, the overall mechanical strength of the material may be lower compared to those employing stronger bonding mechanisms [51]. Despite this, the ability to fine-tune self-healing properties across various temperature ranges and material compositions underscores the value of π–π stacking in designing advanced smart materials.

10.3 PHYSICAL INTERACTIONS

Polymeric materials possess the capability to restore their properties through physical interactions (Figure 10.2) [4]. Following damage, entropy-driven molecular diffusion can initiate the healing process with activation of an external stimulus such as heat [19, 47]. Heating causes the polymer chains in the material to diffuse to form new entanglements and repair the

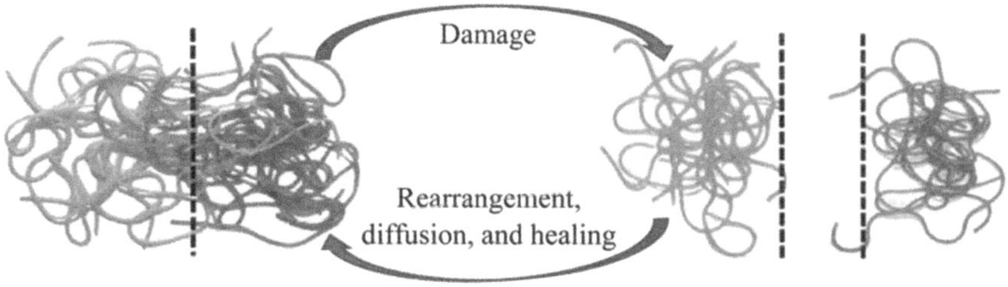

Figure 10.2 Schematic representation of physical self-healing process. **(Reproduced with permission from Ref. [4].)**

cracks. Rahmathullah and Palmese reported multiple self-healing capabilities of epoxy-amine thermoset materials when exposed to temperatures of 185°C, which enhanced molecular diffusion [52]. Similarly, Yamaguchi et al. reported the development of a soft polyurethane (PU) gel by employing molecular diffusion and the entanglement of dangling chains [53, 54].

Both thermoplastic and thermosetting polymers can self-heal through molecular diffusion. The molecules of thermoplastic polymers become mobile at temperatures above their T_g and are therefore more suitable for self-healing *via* molecular interdiffusion [55]. Molecular diffusion occurs between the two damaged parts of a thermoplastic polymer when brought into contact at a temperature above their T_g, leading to the healing of the crack and restoration of mechanical strength [56]. Experimental data on healing conditions suggest that polymers generally heal in a vacuum or at atmospheric pressure, with a range of temperatures from approximately −50°C to +100°C, over varying timeframes from minutes to years [55]. Nevertheless, the T_g is relatively high (above 100°C) in some thermoplastic polymers such as PMMA, and so they require elevated temperatures to initiate self-healing. Thermoset polymers can also exhibit self-healing properties through physical interactions. At either ambient or elevated temperatures, damage in the polymer network can be repaired by chain slippage or interdiffusion of dangling chains. However, thermosets encounter greater challenges compared to thermoplastics due to their cross-linked structure, which creates irreversible chemical bonds. As a result, thermoset polymers have a high T_g, and achieving sufficient molecular mobility often requires heating to very high temperatures. This can lead to molecular degradation and loss of material stability [4].

10.4 SHAPE-MEMORY EFFECTS

Shape-memory materials exhibit a remarkable capability to return to their original shape when subjected to specific stimuli, such as light, heat, or other environmental triggers [57]. This distinctive property is highly beneficial for closing cracks and restoring the structural integrity of composites, making it a valuable mechanism in self-healing materials. These materials can revert to their original form even after undergoing significant, seemingly irreversible deformation on exposure to stimuli such as electrical, temperature or magnetic fields, solutions, or light [58]. This dual functionality, which combines self-healing with shape-memory properties, has attracted considerable interest from researchers, offering promising applications in various fields. The necessity to manually bring damaged parts together to initiate the healing process is one of the primary challenges in developing self-healing materials. Incorporating shape-memory materials into composites enables autonomous healing by utilizing the shape-memory effect to close cracks and restore structural integrity (Figure 10.3). When a crack forms, the shape-memory material can be activated by the appropriate stimulus, causing it to revert to its original shape and seal the crack. This restoration process is not only efficient but also repeatable, allowing the material to heal multiple times.

The use of shape-memory effects in self-healing composites offers several advantages. Shape-memory materials can target specific damaged areas with precision, ensuring effective healing. These materials can be activated multiple times, allowing for repeated healing cycles. Shape-memory materials can be integrated into various composite systems, enhancing their versatility and applicability across different industries. Despite their potential, the use of shape-memory effects in self-healing composites presents some challenges. The need for specific stimuli, such as heat or light, may limit the practical application in environments where controlling these factors is difficult. Secondly, ensuring compatibility between the shape-memory material

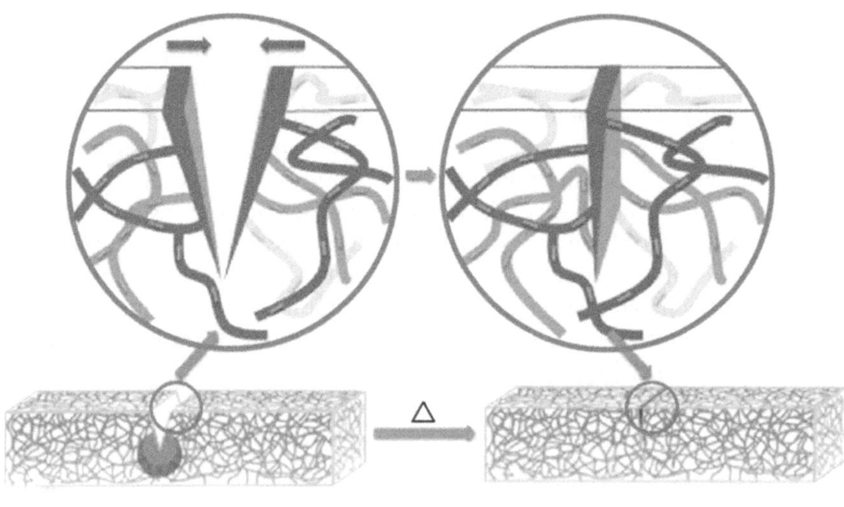

Figure 10.3 Schematic representation of shape-memory assisted self-healing.

Reproduced from Ref [61] with permission from the American Chemical Society, 2013.

and the composite matrix is crucial for maintaining overall material performance. Thirdly, balancing the mechanical properties of the composite with the self-healing capabilities of the shape-memory material requires careful design and optimization. Shape memory-assisted self-healing (SMASH) materials address this issue by automatically closing cracks or damaged areas [59]. Shape-memory materials are programmed to retain a shorter, pre-set shape, which exerts a contracting force on the separated parts, drawing them back into their original position. This ability, which leverages shape-memory properties, has spurred a surge in research aimed at developing self-healing materials. In 2006, Cheng et al. patented a tribological surface capable of self-healing where scratched surfaces were fully restored after being heated to 150°C and then cooled [60]. Similarly, Luo and Mather incorporated electro-spun thermoplastic poly(ε-caprolactone) fibers to develop a SMASH coating into a shape-memory epoxy matrix, which nearly restored the coating's corrosion resistance and healed all the structural damages [61]. Additionally, Rodriguez et al. achieved remarkable self-healing in films with nearly complete recovery by creating a SMASH system by blending a cross-linked poly(ε-caprolactone) network with linear poly(ε-caprolactone) [62]. For shape-memory polymers, the mechanism involves a phase transition between different polymer states. These polymers are typically designed with a network of cross-linked chains that can alternate between a rigid, fixed shape and a flexible, temporary one. When the material is deformed, it is cooled below its transition temperature to lock in the temporary shape [4]. Upon reheating above the transition temperature, the polymer chains regain mobility and return to their original configuration, effectively closing any cracks or voids that have developed. In contrast, shape-memory alloys rely on a crystalline phase transformation. SMAs, such as nickel-titanium (Nitinol), can switch between a martensitic phase (low-temperature phase) and an austenitic phase (high-temperature phase) [63]. When deformed in the martensitic state, the alloy retains the temporary shape until it is heated to the transition temperature, at which point it transforms back to the austenitic phase and recovers its original shape. Ongoing research aims to address these challenges by developing shape-memory materials with lower activation energies, improving the integration techniques with composite matrices, and exploring new stimuli-responsive mechanisms. Advances in this field will enhance the efficiency, reliability, and applicability of shape-memory effects in self-healing composites, paving the way for innovative solutions in materials science and engineering.

10.5 EXTRINSIC SELF-HEALING

Extrinsic self-healing systems function by incorporating microcapsules or vascular networks as external healing agents into a material [64]. These agents are designed to be released when the material sustains damage, initiating a repair process. Extrinsic self-healing mechanisms can be classified into capsule-based self-healing and vascular network self-healing [4]. In capsule-based self-healing, microcapsules are filled with healing agents, which rupture upon damage, and in vascular network self-healing, a more complex system of interconnected channels mimics biological vascular systems, allowing the continuous delivery of healing agents to the affected area. Each approach provides a unique method for autonomous repair, enhancing the longevity and reliability of the material in various applications, such as coatings, structural components, and smart materials.

10.5.1 Capsule-based healing systems

Capsule-based healing systems represent a pioneering approach in the realm of self-healing composites, leveraging the encapsulation of healing agents within microcapsules embedded in the composite matrix [65]. These systems are designed to autonomously repair damage by releasing healing agents when the capsules rupture upon experiencing stress or cracks. This mechanism offers a reliable and efficient means to restore the material's integrity without the need for external intervention [65].

The fundamental principle behind capsule-based healing systems involves embedding microcapsules filled with healing agents into the composite material [66]. These microcapsules are strategically distributed throughout the matrix to ensure that any damage occurring within the composite will likely intersect with one or more capsules. The capsules are designed to remain intact during normal use but to break open when subjected to the mechanical stresses associated with damage, such as cracking or impact [67]. When damage occurs, the mechanical force causes the microcapsules to rupture, releasing the healing agents contained within. These agents typically consist of a resin and a catalyst that react upon contact, initiating a polymerization process that fills and seals the crack. The reaction between the released agents and the surrounding material forms a new, solid bond that restores the structural integrity of the composite. This self-healing process not only mitigates the damage but also prevents further propagation of cracks, enhancing the durability and lifespan of the material. Capsule-based healing systems have the ability to provide targeted and localized repair. Since the healing agents are released precisely at the site of damage, the repair process is highly efficient and effective [66, 67]. Moreover, the encapsulation technique allows for the storage of reactive agents in a stable form, ensuring that they remain inert until needed, thereby maintaining the attributes of the composite. However, the implementation of capsule-based healing systems is not without challenges. One significant issue is ensuring the even distribution of microcapsules throughout the composite matrix [68]. Uneven distribution can lead to areas within the material that are less protected or entirely unprotected by the healing mechanism. Additionally, the size and wall thickness of the microcapsules must be carefully controlled.

Capsules that are too large may weaken the composite, while those that are too small may not contain sufficient healing agents to effectively repair larger cracks. Another challenge is the potential depletion of healing agents after repeated damage [68]. Once the microcapsules in a given area rupture and release their contents, that specific region of the composite may no longer be capable of undergoing self-repair until new capsules are introduced, which could be a limitation in applications requiring long-term durability and multiple repair cycles. More sophisticated microencapsulation techniques are being developed to improve the materials used for both the capsules and the healing agents and to explore the use of multi-capsule systems that contain different types of healing agents for enhanced repair capabilities [69]. Advances in these areas will further enhance the efficiency, reliability, and practicality of capsule-based healing systems, making them an increasingly viable solution for extending the life and performance of composite materials across multiple industries, including civil engineering, automotive, and aerospace. Using this method, White et al. reported the first successful autonomic self-healing technology and achieved a 75% recovery in toughness post-damage [70]. Brown et al. highlighted the effect of size and concentration of microcapsules in healing performance by reporting a 90% recovery in fracture toughness [71].

10.5.2 Vascular healing systems

Vascular healing systems in self-healing composites draw inspiration from the circulatory systems of living organisms, utilizing networks of microchannels filled with healing agents to autonomously repair damage [72]. This approach offers a dynamic and efficient method for restoring the structural integrity of materials by delivering healing agents directly to the affected areas, much like how blood vessels transport nutrients and repair cells to injured tissues in biological systems [73]. The core principle of vascular healing systems involves embedding a network of interconnected microchannels within the composite material, as shown in Figure 10.4.

These channels are pre-filled with healing agents, which typically consist of a resin and a catalyst or other reactive substances. The design of the microchannel network is critical, as it must ensure thorough coverage of the material, allowing healing agents to be transported to any potential damage site effectively. When the composite material incurs damage, such as a crack or fracture, the microchannels intersected by the damage act as conduits for the healing agents [72, 73]. The mechanical force that causes the damage also opens the microchannels, allowing the healing agents to flow out. These agents then migrate toward the damaged area, driven by capillary action or pressure differentials within the system. Upon reaching the damage site, the healing agents mix and react, forming a polymer that seals the crack and restores the material's mechanical properties. One of the primary advantages of vascular healing systems is their ability to continuously supply

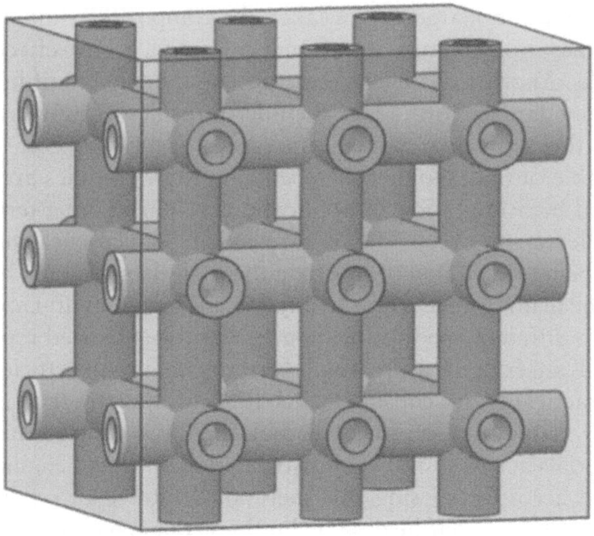

Figure 10.4 A diagram illustrating the microvascular healing concept.

(Reproduced from Ref. [4].)

healing agents to the damaged area until the repair is complete. This continuous flow ensures a thorough filling of the crack and a robust repair, which can significantly enhance the durability and longevity of the material. Additionally, the vascular system can potentially be refilled with new healing agents, allowing for repeated cycles of damage and repair, thereby extending the life of the composite. The effectiveness of vascular healing systems hinges on several key factors, including the design and fabrication of the microchannel network, the selection of healing agents, and the mechanism for initiating the healing process [74]. The microchannels must be carefully engineered to balance structural integrity with functionality; they should be small enough to avoid compromising the composite's mechanical strength but sufficiently large to facilitate the flow of healing agents. Advanced manufacturing techniques, such as 3D printing and microfluidics, are often employed to create these intricate networks within the composite materials. Despite their promising potential, vascular healing systems face several challenges. One significant issue is the potential of the healing agents to cure prematurely or become viscous, which can impede their flow through the microchannels. Researchers are addressing this by developing healing agents with tailored viscosities and reaction kinetics that remain fluid until they reach the damage site. Another challenge is ensuring the robustness of the microchannel network during the composite's fabrication and use, as any pre-existing flaws in the channels can reduce the system's effectiveness [75, 76]. Moreover, integrating vascular healing systems into large-scale manufacturing processes requires addressing scalability and cost-effectiveness. The complexity of embedding microchannels within composites on an industrial scale can be a barrier, necessitating innovations in fabrication techniques and materials. Ongoing research aims to optimize the design and materials used in vascular healing systems, enhance the properties of healing agents, and develop efficient methods for integrating these systems into various composite applications [72, 73]. Advances in these areas will further solidify the role of vascular healing systems in extending the lifespan and reliability of materials used in critical industries such as aerospace, automotive, civil engineering, and beyond. The development of these systems represents a significant step toward more sustainable and resilient materials in engineering and technology.

10.6 APPLICATIONS OF SELF-HEALING COMPOSITES

Self-healing polymers, renowned for their ability to autonomously repair damage, are increasingly finding diverse applications across multiple fields, enhancing durability and extending the lifespan of various products (Figure 10.5) [77]. Their unique attributes make them particularly valuable in industries where material integrity and longevity are crucial. Self-healing polymers are prominently used in consumer electronics and automotive industries [78]. For instance, self-healing protective coatings are applied to smartphones and tablets, allowing them to effectively address surface damage [4].

Similarly, in the high-end automotive sector, self-healing coatings can repair surface damage caused by road debris or parking incidents [77]. This not only reduces maintenance costs but also helps in preserving the vehicle's resale value. In aerospace, self-healing materials play a critical role in maintaining the structural integrity of aircraft and spacecraft. These materials can repair severe damage, such as that caused by bird strikes or abrasion during flight, thereby enhancing safety and reducing maintenance expenses. This application is vital for ensuring the longevity and reliability of aerospace structures. These materials help maintain the functionality and aesthetics of vehicle parts by autonomously repairing minor scratches and dents, thus reducing the need for frequent repairs and replacements. In the medical field, self-healing polymers are used in implants and prosthetics [79]. These materials, such as self-healing hydrogels and biocompatible compounds, can repair minor damage or wear, improving the durability and functionality of medical devices, which is crucial for patient health and device performance. These materials are widely used in tissue engineering, implantable medical devices, and drug delivery systems. Their capacity to repair damage within the human body greatly enhances the safety and efficiency of biomedical applications [79, 80]. Additionally, self healing polymers are utilized in packaging materials to safeguard delicate products during transport and storage [81]. These materials can repair minor

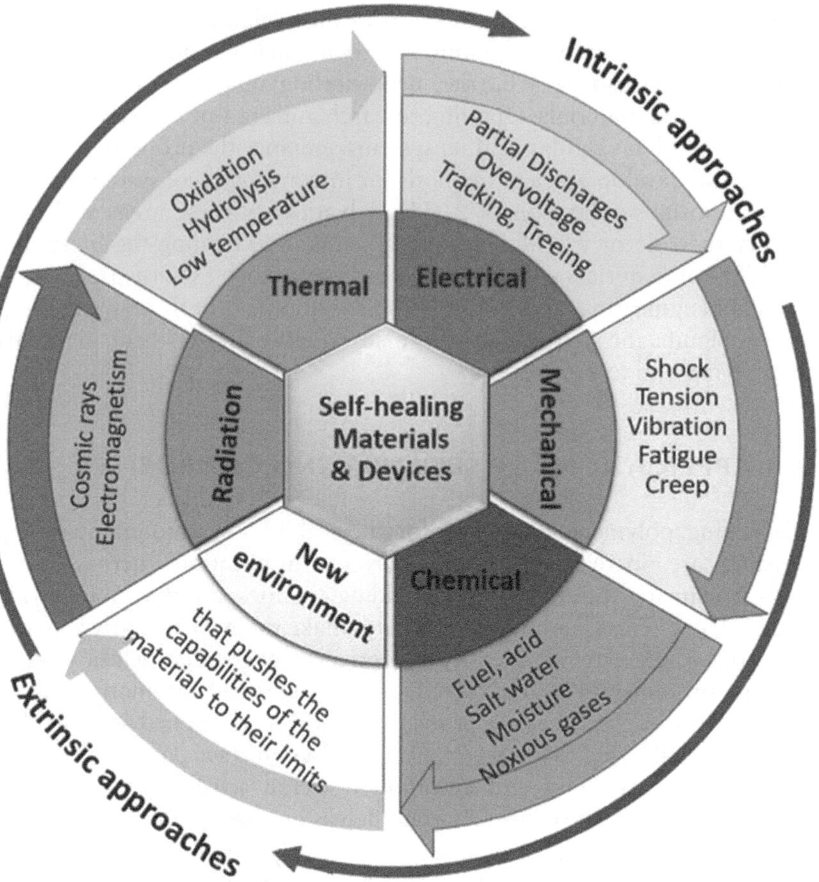

Figure 10.5 Possible applications of self-healing materials and devices in diverse extreme conditions.

(Figure reproduced with permission from Ref. [77].)

punctures or tears, ensuring the integrity of the package and the safety of its contents. Self-healing polymers are incorporated into construction materials, such as concrete and coatings [82]. Self-healing concrete, for example, contains embedded microcapsules or fibers that release healing agents to repair cracks. This innovation helps maintain the structural integrity of buildings and infrastructure, reducing repair costs and extending the lifespan of these essential structures [82]. Incorporating these materials into sealant concrete enables structural elements to self-repair small cracks and fissures, thus preserving durability and long-term functionality. In various applications, self-healing polymers are used to create coatings and protective films that maintain the surface integrity of products [83]. These coatings can repair scratches and abrasions, preserving the aesthetic and functional properties of the coated surfaces. In the textile industry, self-healing polymers are incorporated into fabrics to repair small tears and holes. This application is particularly beneficial for performance wear and outdoor gear, where durability and resilience are essential. Overall, the practical application of self-healing materials spans a wide range of consumer products and industries, offering innovative solutions that improve durability, safety, and performance. Furthermore, the self-healing properties effectively mend microcracks and damage in electrode materials, ensuring the maintenance of battery efficiency and performance over numerous charge-discharge cycles [84]. With the increasing demand for sustainable and long-lasting solutions in energy applications, integrating self-healing materials presents significant potential for improving the reliability and lifespan of energy storage systems, paving the way for future advancements in this domain.

10.7 FUTURE PERSPECTIVES

In recent years, significant research has been devoted to the development of self-healing materials, resulting in notable progress in high-performance polymers and composites with both intrinsic and extrinsic self-repairing abilities. Numerous self-healing mechanisms and strategies have been introduced to meet the needs of practical applications [3, 4]. However, for these materials to transition from experimental concepts to real-world applications, several technical hurdles remain to be addressed. A key challenge for both intrinsic and extrinsic systems is the development of scalable, cost-effective, and high-throughput manufacturing techniques for large-scale production, which remains a major focus of research. Despite this, most studies have focused on the design of self-healing composites for specific functions rather than addressing the challenges of commercialization and mass production. Improving the properties of virgin self-healing materials is critical, as they often perform worse than traditional materials. Additionally, the ability to apply the self-healing function under diverse conditions such as temperature, pressure, humidity, light exposure, and vacuum environments is complex and requires careful design to ensure both long-term durability and efficient healing throughout the material's lifespan. Despite these challenges, ongoing research is expected to overcome these barriers, leading to tailored self-healing materials for various applications. Among extrinsic systems, capsule-based self-healing technologies have seen advancements in capsule durability, healing mechanisms, and types of encapsulated agents [4]. However, a key limitation is that these systems typically allow for only a single healing event. Future work should focus on overcoming this constraint by developing new capsule architectures or integrating shape-memory effects, allowing multiple healing cycles. Research into extrinsic systems has traditionally centered on restoring mechanical properties, but this concept has been expanded to develop multi-functional smart materials, including anti-corrosion coatings, self-healing conductive films, thermosensitive composites, sensors, and wear-resistant, self-lubricating materials [82]. This approach could be further broadened by incorporating bioactive chemicals like antimicrobials or antifouling agents, creating functional coatings for diverse applications. For instance, in civil engineering, self-healing asphalts and concretes with embedded rejuvenator capsules could significantly benefit regions vulnerable to extreme weather and climate change [85]. A promising avenue is the combination of self-healing with self-sensing and shape-memory technologies, which could facilitate autonomous repair in remote or hard-to-reach areas. Over the past decade, microvascular self-healing structures have progressed from simple 2D pipettes to complex 3D networks, improving both the flow of healing agents and the mechanical strength of materials [74]. However, further research is needed to deepen our understanding of the fluid dynamics of healing agents, the behavior of sacrificial vascular networks, and the effects of parameters such as vascular size, shape, wall thickness, and 3D printing speed on the efficiency of self-healing composites. Advances in intrinsic self-healing technologies—such as DA and retro-Diels-Alder (rDA) reactions, ionomer-based systems, dynamic bonding, and thermoplastic/thermoset blends—have gained attention [21, 23]. Nonetheless, most of this work has been limited to thermoplastics, which suffer from lower strength, stiffness, and T_g, making them unsuitable for heavy-duty applications. Still, ionomers have shown potential for rapid crack closure and energy dissipation in high-impact environments. As the focus shifts toward autonomous healing, future intrinsic systems should either eliminate the need for external stimuli or rely only on easily accessible triggers, such as room temperature or UV light from the sun. While tapered double cantilever beam (TDCB) specimens are commonly used to evaluate extrinsic systems, a universal testing standard for intrinsic self-healing materials has yet to be developed [86]. Establishing such a standard is vital for comparative studies, allowing for the accurate assessment of self-healing capabilities across different materials. Therefore, future research must prioritize creating standardized evaluation methods for intrinsic self-healing systems.

10.8 CONCLUSION

Self-healing composites represent a pioneering advancement in materials science, offering the ability to autonomously repair damage, thus extending the lifespan of various materials, drawing inspiration from nature. Self-healing composites incorporating intrinsic and extrinsic healing mechanisms that allow the restoration of mechanical strength and functional integrity after sustaining damage. Recent research has focused on improving these materials' mechanical properties and healing efficiencies, with an emphasis on balancing strength and healing capacity. As we move toward more sustainable technologies, the role of self-healing composites in reducing waste and enhancing the longevity of materials will be crucial. Continued innovation in molecular design, cross-linking strategies, and integration with smart systems will further enhance their application potential. Future research will likely explore the development of multi-functional self-healing materials that not only repair themselves but also offer other advanced features, such as shape memory or enhanced conductivity. In this context, self-healing composites are poised to become integral to the future of durable, high-performance materials across various sectors.

REFERENCES

1. Li, H., Aguirre-Villegas, H.A., Allen, R.D., Bai, X., Benson, C.H., Beckham, G.T., Bradshaw, S.L., Brown, J.L., Brown, R.C., Cecon, V.S. and Curley, J.B., 2022. Expanding plastics recycling technologies: Chemical aspects, technology status and challenges. *Green Chemistry*, 24(23), pp. 8899–9002.
2. Kibria, M.G., Masuk, N.I., Safayet, R., Nguyen, H.Q. and Mourshed, M., 2023. Plastic waste: Challenges and opportunities to mitigate pollution and effective management. *International Journal of Environmental Research*, 17(1), p. 20.
3. Srinivas, M., Yelamasetti, B., Vardhan, T.V. and Mohammed, R., 2021. A critical review on self-healing composites. *Materials Today: Proceedings*, 46, pp. 890–895.
4. Islam, S. and Bhat, G., 2021. Progress and challenges in self-healing composite materials. *Materials Advances*, 2(6), pp. 1896–1926.
5. Cremaldi, J.C. and Bhushan, B., 2018. Bioinspired self-healing materials: Lessons from nature. *Beilstein Journal of Nanotechnology*, 9(1), pp. 907–935.
6. Wen, N., Song, T., Ji, Z., Jiang, D., Wu, Z., Wang, Y. and Guo, Z., 2021. Recent advancements in self-healing materials: Mechanicals, performances and features. *Reactive and Functional Polymers*, 168, p. 105041.
7. Döhler, D., Michael, P. and Binder, W., 2013. Principles of self-healing polymers. *Self-Healing Polymers: From Principles to Applications*, pp. 5–60. DOI:10.1002/9783527670185
8. Syrett, J.A., Becer, C.R. and Haddleton, D.M., 2010. Self-healing and self-mendable polymers. *Polymer Chemistry*, 1(7), pp. 978–987.
9. Yang, Y., Ding, X. and Urban, M.W., 2015. Chemical and physical aspects of self-healing materials. *Progress in Polymer Science*, 49, pp. 34–59.
10. Nevejans, S., Ballard, N., Miranda, J.I., Reck, B. and Asua, J.M., 2016. The underlying mechanisms for self-healing of poly (disulfide) s. *Physical Chemistry Chemical Physics*, 18(39), pp. 27577–27583.
11. Zheng, P. and McCarthy, T.J., 2012. A surprise from 1954: Siloxane equilibration is a simple, robust, and obvious polymer self-healing mechanism. *Journal of the American Chemical Society*, 134(4), pp. 2024–2027.
12. Amamoto, Y., Kamada, J., Otsuka, H., Takahara, A. and Matyjaszewski, K., 2011. Repeatable photoinduced self-healing of covalently cross-linked polymers through reshuffling of trithiocarbonate units. *Angewandte Chemie International Edition*, 50(7), pp. 1660–1663.

13. Otsuka, H., Nagano, S., Kobashi, Y., Maeda, T. and Takahara, A., 2010. A dynamic covalent polymer driven by disulfide metathesis under photoirradiation. *Chemical Communications*, 46(7), pp. 1150–1152.
14. Deng, G., Tang, C., Li, F., Jiang, H. and Chen, Y., 2010. Covalent cross-linked polymer gels with reversible sol– gel transition and self-healing properties. *Macromolecules*, 43(3), pp. 1191–1194.
15. Otsuka, H., Aotani, K., Higaki, Y., Amamoto, Y. and Takahara, A., 2007. Thermal reorganization and molecular weight control of dynamic covalent polymers containing alkoxyamines in their main chains. *Macromolecules*, 40(5), pp. 1429–1434.
16. Yuan, C.E., Rong, M.Z. and Zhang, M.Q., 2014. Self-healing polyurethane elastomer with thermally reversible alkoxyamines as crosslinkages. *Polymer*, 55, pp. 1782–1791.
17. Imato, K., Nishihara, M., Kanehara, T., Amamoto, Y., Takahara, A. and Otsuka, H., 2012. Self-healing of chemical gels cross-linked by diarylbibenzofuranone-based trigger-free dynamic covalent bonds at room temperature. *Angewandte Chemie*, 51(5), pp. 1138–1142.
18. Froimowicz, P., Frey, H. and Landfester, K., 2011. Towards the generation of self-healing materials by means of a reversible photo-induced approach. *Macromolecular Rapid Communications*, 32(5), pp. 468–473.
19. Chen, X., Dam, M.A., Ono, K., Mal, A., Shen, H., Nutt, S.R., Sheran, K. and Wudl, F., 2002. A thermally re-mendable cross-linked polymeric material. *Science*, 295(5560), pp. 1698–1702.
20. Rekondo, A., Martin, R., de Luzuriaga, A.R., Cabañero, G., Grande, H.J. and Odriozola, I., 2014. Catalyst-free room-temperature self-healing elastomers based on aromatic disulfide metathesis. *Materials Horizons*, 1(2), pp. 237–240.
21. Garcia, S.J., 2014. Effect of polymer architecture on the intrinsic self-healing character of polymers. *European Polymer Journal*, 53, pp. 118–125.
22. Amamoto, Y., Otsuka, H., Takahara, A. and Matyjaszewski, K., 2012. Self-healing of covalently cross-linked polymers by reshuffling thiuram disulfide moieties in air under visible light. *Advanced Materials (Deerfield Beach, Fla.)*, 24(29), pp. 3975–3980.
23. Tsarevsky, N.V. and Matyjaszewski, K., 2002. Reversible redox cleavage/coupling of polystyrene with disulfide or thiol groups prepared by atom transfer radical polymerization. *Macromolecules*, 35(24), pp. 9009–9014.
24. Scheiner, M., Dickens, T.J. and Okoli, O., 2016. Progress towards self-healing polymers for composite structural applications. *Polymer*, 83, pp. 260–282.
25. Higaki, Y., Otsuka, H. and Takahara, A., 2006. A thermodynamic polymer cross-linking system based on radically exchangeable covalent bonds. *Macromolecules*, 39(6), pp. 2121–2125.
26. Kamada, J., Koynov, K., Corten, C., Juhari, A., Yoon, J.A., Urban, M.W., Balazs, A.C. and Matyjaszewski, K., 2010. Redox responsive behavior of thiol/disulfide-functionalized star polymers synthesized via atom transfer radical polymerization. *Macromolecules*, 43(9), pp. 4133–4139.
27. Urban, M.W., 2012. The chemistry of self-healing. *Nature Chemistry*, 4(2), pp. 80–82.
28. Ghosh, B., Chellappan, K.V. and Urban, M.W., 2011. Self-healing inside a scratch of oxetane-substituted chitosan-polyurethane (OXE-CHI-PUR) networks. *Journal of Materials Chemistry*, 21(38), pp. 14473–14486.

29. Burattini, S., Colquhoun, H.M., Greenland, B.W. and Hayes, W., 2009. A novel self-healing supramolecular polymer system. *Faraday Discussions*, 143, pp. 251–264.
30. Yang, Y. and Urban, M.W., 2018. Self-healing of polymers *via* supramolecular chemistry. *Advanced Materials Interfaces*, 5(17), p. 1800384.
31. Tee, B.C., Wang, C., Allen, R. and Bao, Z., 2012. An electrically and mechanically self-healing composite with pressure-and flexion-sensitive properties for electronic skin applications. *Nature Nanotechnology*, 7(12), pp. 825–832.
32. Vega, J.M., Van der Zwaag, S. and Garcia Espallargas, S.J., 2013. The role of clusters on the healing efficiency of a modified Zn based ionomer. In *ICSHM 2013: Proceedings of the 4th International Conference on Self-Healing Materials*, Ghent, Belgium, June 16–20, 2013. Ghent University; Delft University of Technology.
33. Hohlbein, N., von Tapavicza, M., Nellesen, A. and Schmidt, A.M., 2013. Self-healing ionomers. In *Self-healing Polymers: From Principles to Applications*, Wiley, pp. 315–334. DOI:10.1002/9783527670185.ch13
34. Tachine, H., Hara, H., Hirasawa, E., Tadano, K., Yano, S. 1993. Dynamic mechanical relaxations of ethylene ionomers. *Macromolecules*, 26(4), pp. 752–757. DOI:doi.org/10.1021/ma00056a029
35. Sijbesma, R.P., Beijer, F.H., Brunsveld, L., Folmer, B.J., Hirschberg, J.K., Lange, R.F., Lowe, J.K. and Meijer, E.W., 1997. Reversible polymers formed from self-complementary monomers using quadruple hydrogen bonding. *Science*, 278(5343), pp. 1601–1604.
36. ARKEMA: 'Reverlink Supramolecular Technology', 2013. https://www.arkema.com/en/
37. Campanella, A., Döhler, D. and Binder, W.H., 2018. Self-healing in supramolecular polymers. *Macromolecular Rapid Communications*, 39(17), p. 1700739.
38. Wool, R. and O'Connor, K.M., 1981. A theory crack healing in polymers. *Journal of Applied Physics*, 52(10), pp. 5953–5963.
39. Bosman, A.W., Sijbesma, R.P. and Meijer, E.W. *Materials Today*, 2004, 7(4), pp. 34–39. DOI:10.1016/S1369-7021(04)00187-7
40. Söntjens, S.H., Sijbesma, R.P., van Genderen, M.H.P. and Meijer, E.W. *Journal of the American Chemical Society*, 2000, 122, pp. 7487–7493. DOI:10.1021/ja000435m
41. Yamauchi, K., Lizotte, J.R., Hercules, D.M., Vergne, M.J. and Long, T.E., 2002. *Journal of the American Chemical Society*, 124, 8599–8604. DOI:10.1021/ja020123e
42. Feldman, K.E., Kade, M.J., de Greef, T.F., Meijer, E.W., Kramer, E.J. and Hawker, C.J., 2008. Polymers with multiple hydrogen-bonded end groups and their blends. *Macromolecules*, 41(13), pp. 4694–4700.
43. de Greef, T.F. and Meijer, E.W., 2008. Supramolecular polymers. *Nature*, 453(7192), pp. 171–173.
44. De Greef, T.F., Smulders, M.M., Wolffs, M., Schenning, A.P., Sijbesma, R.P. and Meijer, E.W., 2009. Supramolecular polymerization. *Chemical Reviews*, 109(11), pp. 5687–5754.
45. Lin, C.B., Lee, S. and Liu, K.S., 1990. Methanol-induced crack healing in poly (methyl methacrylate). *Polymer Engineering & Science*, 30(21), pp. 1399–1406.
46. Hager, M.D., Greil, P., Leyens, C., Van Der Zwaag, S. and Schubert, U.S., 2010. Self-healing materials. *Advanced Materials*, 22(47), pp. 5424–5430.

47. Zhang, P. and Li, G., 2016. Advances in healing-on-demand polymers and polymer composites. *Progress in Polymer Science*, 57, pp. 32–63. DOI: 10.1016/j.progpolymsci.2015.11.005
48. Li, C.H. and Zuo, J.L., 2020. Self-healing polymers based on coordination bonds. *Advanced Materials*, 32(27), p. 1903762.
49. Yang, Y. and Urban, M.W., 2013. Self-healing polymeric materials. *Chemical Society Reviews*, 42(17), pp. 7446–7467.
50. Burattini, S., et al., 2010. *Journal of the American Chemical Society*, 132, pp. 12051–12058. DOI:10.1021/ja104446r
51. Burattini, S., Colquhoun, H.M., Fox, J.D., Friedmann, D., Greenland, B.W., Harris, P.J., Hayes, W., Mackay, M.E. and Rowan, S.J., 2009. A self-repairing, supramolecular polymer system: Healability as a consequence of donor–acceptor π–π stacking interactions. *Chemical Communications*, 44, pp. 6717–6719.
52. Rahmathullah, M.A.M. and Palmese, G.R., 2009. Crack-healing behavior of epoxy–amine thermosets. *Journal of Applied Polymer Science*, 113, pp. 2191–2201. DOI:10.1002/app.30152
53. Yamaguchi, M., Ono, S. and Okamoto, K., 2009. *Materials Science and Engineering B*, 162, pp. 189–194. DOI:10.1016/j.mseb.2009.04.006
54. Yamaguchi, M., Ono, S. and Terano, M., 2007. Self-repairing property of polymer network with dangling chains. *Advanced Materials Letters*, 61, pp. 1396–1399. DOI:10.1016/j.matlet.2006.07.039
55. Van der Zwaag, S., Van Dijk, N.H., Jonkers, H.M., Mookhoek, S.D. and Sloof, W.G., 2009. Self-healing behaviour in man-made engineering materials: Bioinspired but taking into account their intrinsic character. *Philosophical Transactions of the Royal Society A: Mathematical, Physical and Engineering Sciences*, 367(1894), pp. 1689–1704.
56. Wu, D.Y., Meure, S. and Solomon, D., 2008. Self-healing polymeric materials: A review of recent developments. *Progress in Polymer Science*, 33(5), pp. 479–522.
57. Corten, C.C. and Urban, M.W., 2009. Repairing polymers using oscillating magnetic field. *Advanced Materials*, 21, pp. 5011–5015. DOI:10.1002/adma.200901940
58. Cortie, M.B. and McDonagh, A.M., 2011. Synthesis and optical properties of hybrid and alloy plasmonic nanoparticles. *Chemical Reviews*, 111, pp. 3713–3735. DOI:10.1021/cr1002529
59. Lai, H.Y., Wang, H.-Q., Lai, J.-C. and Li, C.-H., 2019. A self-healing and shape memory polymer that functions at body temperature. *Molecules*, 24, pp. 1–12. DOI:10.3390/molecules24183224
60. Cheng, Y.T., Ni, W., Lukitsch, M.J., Weiner, A.M. and Grummon, D.S., 2006. Motors Liquidation Co and Michigan State University MSU. *Self-healing tribological surfaces*, US Pat. 7, 060,140.
61. Luo, X. and Mather, P.T.,2013. Shape memory assisted self-healing coating. *ACS Macro Letters*, 2, pp. 152–156. DOI:10.1021/mz400017x
62. Rodriguez, E.D., Luo, X. and Mather, P.T.,2011. Linear/network poly (ε-caprolactone) blends exhibiting shape memory assisted self-healing (SMASH). *ACS Applied Materials & Interfaces*, 3, pp. 152–161. DOI:10.1021/am101012c

63. Gallardo Fuentes, J.M., Gümpel, P. and Strittmatter, J., 2002. Phase change behavior of nitinol shape memory alloys. *Advanced Engineering Materials*, 4(7), pp. 437–452.
64. Urdl, K., Kandelbauer, A., Kern, W., Müller, U., Thebault, M. and Zikulnig-Rusch, E., 2017. Self-healing of densely crosslinked thermoset polymers—A critical review. *Progress in Organic Coatings | Journal*, 104, pp. 232–249. DOI:10.1016/j.porgcoat.2016.11.010
65. Mohonee, V.K., Goh, K.L., Mishnaevsky Jr, L. and Pasbakhsh, P., 2021. Capsule based self-healing composites: New insights on mechanical behaviour based on finite element analysis. *Computational Materials Science*, 192, p. 110203.
66. Kosarli, M., Bekas, D., Tsirka, K. and Paipetis, A.S., 2020. Capsule-based self-healing polymers and composites. In *Self-healing Polymer-Based Systems*. Elsevier, pp. 259–278.
67. Ilyaei, S., Sourki, R. and Akbari, Y.H.A., 2021. Capsule-based healing systems in composite materials: A review. *Critical Reviews in Solid State and Materials Sciences*, 46(6), pp. 491–531.
68. Zhu, D.Y., Rong, M.Z. and Zhang, M.Q., 2015. Self-healing polymeric materials based on microencapsulated healing agents: From design to preparation. *Progress in Polymer Science*, 49, pp. 175–220.
69. Sadabadi, H., Allahkaram, S.R., Kordijazi, A. and Rohatgi, P.K., 2022. Self-healing coatings loaded by nano/microcapsules: A review. *Protection of Metals and Physical Chemistry of Surfaces*, 58(2), pp. 287–307.
70. White, S.R., et al., 2001. Autonomic healing of polymer composites. *Nature*, 409, pp. 794–797. DOI:10.1038/35057232
71. Brown, E.N., Sottos, N.R. and White, S.R., 2002. Fracture testing of a self-healing polymer composite. *Experimental Mechanics*, 42, pp. 372–379.
72. Williams, H.R., Trask, R.S. and Bond, I.P., 2008. Self-healing sandwich panels: Restoration of compressive strength after impact. *Composites Science and Technology*, 68(15–16), pp. 3171–3177.
73. Williams, H.R., Trask, R.S., Knights, A.C., Williams, E.R. and Bond, I.P., 2008. *Journal of the Royal Society Interface*, 5, pp. 735–747. DOI:10.1098/rsif.2007.1251
74. Cuvellier, A., Torre-Muruzabal, A., Van Assche, G., De Clerck, K. and Rahier, H., 2017. Selection of healing agents for a vascular self-healing application. *Polymer Testing*, 62, pp. 302–310.
75. Shields, Y., De Belie, N., Jefferson, A. and Van Tittelboom, K., 2021. A review of vascular networks for self-healing applications. *Smart Materials and Structures*, 30(6), p. 063001.
76. Lee, M.W., An, S., Yoon, S.S. and Yarin, A.L., 2018. Advances in self-healing materials based on vascular networks with mechanical self-repair characteristics. *Advances in Colloid and Interface Science*, 252, pp. 21–37.
77. Ekeocha, J., Ellingford, C., Pan, M., Wemyss, A.M., Bowen, C. and Wan, C., 2021. Challenges and opportunities of self-healing polymers and devices for extreme and hostile environments. *Advanced Materials*, 33(33), p. 2008052.
78. Sabet, M., 2024. Unveiling advanced self-healing mechanisms in graphene polymer composites for next-generation applications in aerospace, automotive, and electronics. *Polymer-Plastics Technology and Materials*, 63(15), pp. 2032–2059.

79. Diba, M., Spaans, S., Ning, K., Ippel, B.D., Yang, F., Loomans, B., Dankers, P.Y. and Leeuwenburgh, S.C., 2018. Self-healing biomaterials: From molecular concepts to clinical applications. *Advanced Materials Interfaces*, 5(17), p. 1800118.
80. Chen, J., Huang, Y., Ma, X. and Lei, Y., 2018. Functional self-healing materials and their potential applications in biomedical engineering. *Advanced Composites and Hybrid Materials*, 1, pp. 94–113.
81. Wang, J., Gao, Q., Zhao, F. and Ju, J., 2024. Repair mechanism and application of self-healing materials for food preservation. *Critical Reviews in Food Science and Nutrition*, 64(30), pp. 11113–11123.
82. Nik Md Noordin Kahar, N.N.F., Osman, A.F., Alosime, E., Arsat, N., Mohammad Azman, N.A., Syamsir, A., Itam, Z. and Abdul Hamid, Z.A., 2021. The versatility of polymeric materials as self-healing agents for various types of applications: A review. *Polymers*, 13(8), p. 1194.
83. Zhang, F., Ju, P., Pan, M., Zhang, D., Huang, Y., Li, G. and Li, X., 2018. Self-healing mechanisms in smart protective coatings: A review. *Corrosion Science*, 144, pp. 74–88.
84. Ezeigwe, E.R., Dong, L., Manjunatha, R., Tan, M., Yan, W. and Zhang, J., 2021. A review of self-healing electrode and electrolyte materials and their mitigating degradation of Lithium batteries. *Nano Energy*, 84, p. 105907.
85. Tabaković, A. and Schlangen, E., 2016. Self-healing technology for asphalt pavements. In *Self-Healing Materials*. Springer Nature, pp. 285–306.
86. Khorasani, S.N. and Neisiany, R.E., 2020. Characterization of self-healing polymeric materials. In *Self-Healing Polymer-Based Systems*. Elsevier, pp. 123–140.

Index

Pages in *italics* refer to figures and pages in **bold** refer to tables.